Contrastivism in Philosophy

Contrastivism can be applied to a variety of problems within philosophy, and as such, it can be coherently seen as a unified movement. This volume brings together state-of-the-art research on the contrastive treatment of philosophical concepts and questions, including knowledge, belief, free will, moral luck, Bayesian confirmation theory, causation, and explanation.

Martijn Blaauw is Assistant Director of the 3TU.Centre for Ethics and Technology and a Senior Research Fellow at the Philosophy Section of Delft University of Technology.

Routledge Studies in Contemporary Philosophy

Contrastivism in Philosophy

Edited by Martijn Blaauw

Routledge
Taylor & Francis Group
NEW YORK   LONDON

First published 2013
by Routledge
711 Third Avenue, New York, NY 10017

Simultaneously published in the UK
by Routledge
2 Park Square, Milton Park, Abingdon, Oxfordshire OX14 4RN

First issued in paperback 2015

*Routledge is an imprint of the Taylor & Francis Group,
an informa business*

Library of Congress Cataloging-in-Publication Data
Contrastivism in philosophy : new perspectives / edited by Martijn
    Blaauw.
        pages cm. — (Routledge studies in contemporary philosophy ; 39)
        Includes bibliographical references and index.
    1. Contrastivism (Philosophy)   I. Blaauw, Martijn, editor of
compilation.
    B809.145.C68 2012
    149—dc23
    2011048593

ISBN 13: 978-1-138-92283-9 (pbk)
ISBN 13: 978-0-415-87860-9 (hbk)

Typeset in Sabon
by IBT Global.

Contents

Figures

Acknowledgments

I would like to thank the authors who contributed to this volume for their chapters and for their cooperative attitude during the process of completing this collection. I am especially grateful to Walter Sinnott-Armstrong, who has been an essential motivation behind this project and who hosted some of the contributors to this volume at a contrastivism workshop at Dartmouth College in 2007. I gratefully acknowledge a VENI Research Grant (2007–2010) awarded to me by the Netherlands Organisation for Scientific Research (NWO), which has enabled me to complete this collection. Finally, I would like to express my gratitude to Felisa Salvago-Keyes and Erica Wetter, my editors at Routledge Publishers, for encouragement and admirable support throughout.

Introduction

Contrastivism in Philosophy

Martijn Blaauw

In recent years, there has been a substantial amount of interest in contrastivist approaches to philosophical concepts. So much so, that one might even speak of a contrastivist *movement* in philosophy. Versions of contrastivism that have recently been defended are contrastive knowledge, contrastive causation, and contrastive explanation.[1] As Walter Sinnott-Armstrong, one of the movement's principal members, has put it recently:

> A spectre is haunting epistemology—the spectre of contrastivism. . . . It is high time that contrastivists should openly, in the face of the whole world, publish their views, their aims, their tendencies, and meet this nursery tale of the spectre of contrastivism with a manifesto of the party itself. (2008, 257)

One idea underlying this collection of essays is that this specter haunts not only epistemology, but philosophy in its entirety. However, another suggestion essential to this volume is that the specter, when properly understood and confronted, might not even *haunt* philosophy but have a rather benign influence instead. Indeed, the contrastivist specter might actually be an angel of accession.

In this introduction, I will do two main things. In the first place, I will try to describe contrastivism as a movement in philosophy, by focusing on the following three questions: (i) What is the nature of contrastivism? (ii) What is the purpose of contrastivism? (iii) Which varieties of contrastivism can be distinguished? In the second place, I will provide a guided tour through the chapters—and different versions of contrastivism—that constitute this volume.

1 CONTRASTIVISM: NATURE AND PURPOSE

What is the *nature* of contrastivism? Sinnott-Armstrong, in his chapter in this volume, has provided a useful description of the view. He writes:

> A contrastivist view of a concept holds that all or some claims using
> that concept are best understood with an extra logical place for a con-
> trast class. (134)

Central to this description of contrastivism is the idea that, for a particular
concept, an extra logical place for a contrast class should be introduced. For
present purposes, think of concepts as expressing relations. The concept of
knowledge expresses a relation between a subject and a proposition; the
concept of causation expresses a relation between cause and effect; the concept of explanation expresses a relation between explanans and explanandum, and so on. Now contrastivizing a concept means changing the adicity
of the relation by adding a relatum for a contrast class.

What is the *purpose* of giving a contrastive account of a particular concept? Why add an extra relatum for a contrast class? There might be several
advantages to contrastivism. For instance, it can be argued that adding a
contrast class has *puzzle-solving potential*. A contrastive account of knowledge and/or belief might, for instance, help solve the puzzle of radical skepticism (see the contributions by Morton and Blaauw in this volume). A
contrastive account of causation might help to "resolve paradoxes as to
whether absences are causal . . . whether events are fragile . . . whether
causation is extensional . . . whether causation is transitive . . . and whether
selection of 'the cause' is objective." (Schaffer 2005a, 299; also see Schaffer's chapter on contrastive causation in this volume). A contrastive account
of moral luck can help solve the problem of moral luck (see Driver's Chapter
9); a contrastive account of deontic modals can solve various puzzling cases
about "ought," "must," and "may" (Snedegar's Chapter 7); and contrastive accounts of free will (Sinnott-Armstrong's Chapter 8) and explanation
(Hitchcock's Chapter 1) have been argued to solve various puzzles as well.

Another reason to defend contrastivism can be *precision*. We don't need
our language to be optimally precise all the time. But sometimes, precision is needed. We could need to specify exactly what someone knows or
believes, what something causes or explains, what someone ought to do,
in what respect someone has free will, or to what extent one's actions are
lucky. A nice example is provided by Walter Sinnott-Armstrong in his chapter on contrastive free will, where he discusses the trial of John Hinckley
and shows that contrastivism—*contrastivist precision*—can actually help
to see through a fallacy put forward by the prosecutors. This indicates
that contrastivism isn't a purely academic exercise—interesting, perhaps,
to philosophers, but deeply irrelevant to all others. Contrastivist precision
can have important real-life applications and implications.

If we now return to the description of contrastivism given by Sinnott-
Armstrong in the quotation above, a couple of questions arise. First, the
quotation speaks about "*claims* using the concept"; here we could ask: can
a contrastive view also be about the concept *itself*? Second, the quotation
speaks about "*all or some* claims using that concept"; here we could ask:

what is the *scope* of a contrastive theory? Third, the quotation speaks about those claims being "best understood" in a contrastive fashion; here we could ask: should a contrastive theory of a particular concept do justice to how we normally use the concept? To this list of questions, I would add the following two: should one determine whether one particular contrast class is the relevant one? And: is there a limit to how many different contrast classes can be introduced? Answering these questions will provide alternative versions of contrastivism that can differ along the following dimensions: attributer contrastivism versus nonattributer contrastivism; universal contrastivism versus particular contrastivism; revisionism versus nonrevisionism; Pyrrhonian versus non-Pyrrhonian contrastivism; and single-contrast versus multicontrast contrastivism. In the next section, I will briefly elucidate these versions of contrastivism.

2   VERSIONS OF CONTRASTIVISM

Let's start with the issue of attributer contrastivism versus nonattributer contrastivism. Here one could argue that contrastivism with respect to a particular concept pertains to attributions of that concept ("attributer contrastivism"), or one could argue that contrastivism with respect to a particular concept pertains to the concept itself ("nonattributer contrastivism"). Also, one could argue that both the concept and attributions of that concept are contrastive in nature. This leads to the question of what the relation between the two versions of contrastivism is. It seems at least plausible that if the concept itself displays contrastivity, attributions of that concept must follow suit. The other way around, however, might be less clear: if attributions of a concept always involve contrasts, is the concept itself contrastive? The answer to this question depends, it seems, on whether one thinks the contrasts are semantic or pragmatic in nature. The essays in this volume, if they take an explicit stand on this, are mostly concerned with attributer contrastivism.

Turning to the distinction between universal versus particular contrastivism, this distinction concerns the scope of the contrastive theory. Take, for example, the concept of causation. If one is a contrastivist about causation, one could either defend that all attributions of causation are to be understood contrastively, or that only *some* attributions of causation are to be understood contrastively. In his contribution to this volume, Adam Morton defends the position that not necessarily all attributions of knowledge are contrastive in nature, leaving open the option of a binary "knows." The other essays in this volume either don't take an explicit stand on this issue or seem to defend a universal type of contrastivism with respect to the concept in question.

As to the distinction between revisionism and nonrevisionism, the problem here is whether the concept in question should *from now on* be understood as being contrastive in nature or *always was* contrastive in nature (even though we perhaps didn't realize this). The contributions to this volume do not explicitly discuss this issue, but the topic is weaved into some of the papers. The contribution of Jonathan Schaffer, for instance, in which he argues that causation is contrastive, comes closest to taking a stand on this issue. Schaffer seems to accept that his version of contrastivism reflects the way in which ordinary speakers of English use the verb "to cause." Also, the chapter on epistemic contrastivism by Adam Morton starts from the idea that the contrastive use of "to know" is embedded in our everyday language. If one decided to be a revisionist about a certain concept and argue that it is contrastive in nature even though nobody thinks it is, this might be justified if replacing the normal, two-place concept with a three-place, contrastive concept can solve problems that would otherwise be left unresolved.

Yet another distinction is that between Pyrrhonian contrastivism and non-Pyrrhonian contrastivism. Contrastivist theories introduce an extra logical place for a contrast class. This contrast class will not be empty: it will contain one (or more) contrastive propositions. But what is, in a given situation, the relevant set of contrastive propositions? How do we determine this? I have previously called this question the "relevance question" (Blaauw 2008b, 473). One can answer the relevance question in three main ways. In the first place, one could argue that one specific contrast class is always the relevant contrast class; this would be a form of "invariantism." In the second place, one could argue that which contrast class is relevant depends on features of the (conversational) context. Sinnott-Armstrong has argued for a third option: one should *suspend belief about which contrast class is relevant*. This position—*Pyrrhonism*—provides an elegant solution to the pressing problem of what makes alternatives relevant.

A final distinction concerns the number of contrast places one might wish to defend with respect to a particular concept. Where contrastive accounts of knowledge and belief, for instance, are standardly defended as being three-place in nature (S knows that p rather than Q; S believes that p rather than Q), Jonathan Schaffer defends a four-place causal relation (c rather than c' causes e rather than e'). Similarly, Walter Sinnott-Armstrong defends that free will is contrastive along two dimensions: *free from x rather than y* and *free to do x rather than y*, also resulting, or so it seems, in a four-place free-will relation.

These different types of contrastivism can, I think, be mixed and matched. Now after having read the essays in this volume, one might well ask (either exasperated or delighted): "*Where does it all end? Can any* concept or attribution of a concept be contrastivized?" Unless one is a non-revisionist, I see no reason to suppose that some concepts are, in principle, excluded from a contrastive treatment. For if one is a non-revisionist, a concept can only be used contrastively if it can be shown that there is linguistic evidence that supports this. But, of course, contrastivizing a concept must have a purpose: it must be able to solve otherwise unsolvable puzzles. Without such puzzle-solving potential, randomly adding contrast classes seems only to complicate rather than enlighten. With all this in mind, let us now turn to the contributions to this volume.

3 THE CONTRIBUTIONS

Contrastive theories of explanation might be where it all began. The chapter by Christopher Hitchcock in this collection, "Contrastive Explanation," provides a discussion of different types of contrastive explanation. He discusses three different dimensions of contrastivism: wide versus narrow scope, contrast in the explanans versus contrast in the explanandum, and contrasting processes versus contrasting outcomes. These dimensions pertain to what goes into the contrast set, where the contrast is located, and what the relation is between processes and outcomes. Hitchcock then goes on to present a picture of explanation, where he takes an explanation "to provide us with information about what the explanandum *depends on*" (19). In other words, an explanation "gives us information about how changes in the conditions described in the explanans would result in differences in the explanandum" (19). Hitchcock then illustrates this with seven different types of explanation. All this culminates in a general schema for explanation which shows why explanation is contrastive: contrasts play the role of specifying which variables are involved in the explanation. For instance, the sentence "Adam ate the apple rather than the pear" contains the variable AE ("possible things that Adam eats"), which is restricted to two possible values (the apple and the pear), where AE takes the value "apple." This schema is then applied to the three different dimensions of explanation.

Causation and explanation are closely related concepts. In Chapter 2, "Causal Contextualisms," Jonathan Schaffer argues for a contrastivist account of causation. Key to his account is that the context sensitivity of causal claims is due not to pragmatic factors but to semantic factors instead. Schaffer starts out by providing three examples that illustrate the context sensitivity of causal claims alongside three examples that illustrate the sensitivity of causal claims to event descriptions. He then considers two main ways to deal with these sensitivities: the invariantist orthodoxy, which holds that the context sensitivity of causal claims is due to pragmatics; and the new contextualism, which holds that the context sensitivity of causal claims arises in part due to the semantics of "to cause." Schaffer provides three arguments against the pragmatic view, thus providing a *prima facie* case for the new contextualism. Schaffer then goes on to argue for a particular version of contextualism: contrastivism, and more specifically, the double-contrast view: c rather than C* causes e rather than E*. Schaffer concludes by considering the question of where the sensitivity of causal claims arises and concludes with a puzzle, hoping that it can be solved by using covert variables, but ultimately leaving open how this might be implemented in a plausible semantics.

Branden Fitelson, in the third chapter of this volume ("Contrastive Bayesianism"), shows how contrastivist thinking arises in some recent applications of Bayesian techniques by examining a number of case studies from the extant literature on Bayesian confirmation theory. Fitelson starts with a discussion of a contrastivist probabilistic account taken from philosophy of science. This brings to the fore questions about (the relations between) Likelihoodism, Bayesianism, "favoring," and "contrastive confirmation." He then goes on to discuss the problem of the irrelevant conjunction. This problem provides a class of examples where evidence E supports H more strongly than E supports H & X—the proposed solution to this problem is then discussed further, leading up to a discussion of the "conjunction fallacy." It becomes apparent that this is a rich field of research where philosophical and empirical work will reinforce each other.

In the fourth chapter, "Contrastive Belief," Martijn Blaauw argues for a contrastive account of the concept of "belief". Attributions of belief, Blaauw argues, are to be understood in ternary, contrastive terms: S believes that p rather than Q. After having presented some examples that seem to indicate that it is natural to interpret belief attributions in a contrastive way, Blaauw goes on to argue that what it means to believe a proposition can be captured in terms of "comparative confidences": S believes that p means that S is more confident that p than q. So it might be the case that one is more confident that it rains rather than snows but isn't more confident that it rains rather than sleets. After having considered various refinements and objections, Blaauw shows how the idea of contrastive belief can provide a new type of solution to the problem of radical skepticism. The essence of the problem of radical skepticism isn't that we have insufficient evidence for our beliefs. It is that our beliefs are simply less heavyweight than we assumed.

Adam Morton, in his "Contrastive Knowledge" (Chapter 5), argues in favor of contrastive knowledge. Specifically, Morton argues for a version of epistemic contrastivism that leaves open the possibility that we can think of a simpler, non-contrastive knowledge relation. The upshot of his chapter is that the way the concept of knowledge functions in our everyday language can be best captured by a contrastive knowledge relation. So how does the concept of knowledge function, according to Morton? First, when we explain someone's actions we often appeal to the tracking relations to their environment, on which people base many of their actions, and, essentially related to this, provide themselves with knowledge. But tracking is by its very nature contrastive, due to the limited cognitive capacities of humans. We can, for instance, predict where the animal is going as long as it hasn't taken certain evasive measures. Or we can explain why it is raining instead of snowing, but not why it is raining instead of sleeting. We need contrasts to be optimally precise. Besides tracking, *evidence* is also an important basic source of knowledge. Central when it comes to evidence are assumptions. When seeking evidence that a coin is fair, for instance, one assumes that it has a constant bias. The role of assumptions in evidence can be accommodated by contrastivism more easily than by other, binary, accounts of knowledge. You know that the coin is fair rather than strongly biased to heads; you don't know that the coin is fair rather than of varying bias. You know that there is a dog in front of you rather than a cat; you don't know that there is a dog in front of you rather than a hallucination. Morton concludes that the reason we attribute knowledge is that we are curious about what aspects of the world creatures have accurate information about to guide their actions. Their actions can then be used as a guide to our own actions. And therefore we need information-streams that are appropriately wide or narrow. Contrastive knowledge attributions can take this desideratum into account.

Justin Snedegar, in his "Contrastive Semantics for Deontic Modals" (Chapter 6), develops a contrastive framework for "ought," in line with philosophers such as Sloman, Jackson, and Cariani. He argues further that this framework can be extended to cover "must" and "may" as well. Key to Snedegar's defense of a contrastive framework for "ought" are two puzzling cases, the reconciliation puzzle and the inheritance puzzle. The reconciliation puzzle is that it can both be true that (i) one ought to F and (ii) one ought not to F. For instance, it can be that one both ought to accept and write the book review and ought not to accept. Snedegar argues that the contrastivity of "ought" can solve this puzzling situation; the inconsistency only arises if one fails to recognize this contrastivity. The second puzzle is based on the Inheritance Principle: if p entails q, then if it ought to be that p, it ought to be that q. But this brings trouble of the following sort. Suppose the following claim is true: "It ought to be that you help the injured stranger." By Inheritance, it follows that "It ought to be that there is an injured stranger and you help him." The latter claim is presumably false: it ought to be that there is *not* an injured stranger! Again, contrastivism about "ought" can solve the puzzle "by pointing out that the contrast has shifted [between the two sentences]" (119). Snedegar goes on to show that similar puzzles arise for "must" and "may," and, after having considered various objections, proposes that contrastivism can solve the puzzles in these cases as well. Finally, Snedegar shows that a contrastive account of "ought," "must," and "may" can be captured in a semantic framework that accommodates all the desired relationships between these modals.

Walter Sinnott-Armstrong, in his contribution "Free Contrastivism" (Chapter 7), builds on his previously defended version of contrastivism— Pyrrhonian contrastivism. A core idea in Sinnott-Armstrong's contrastivism is that reasons are contrastive in nature. Reasons to believe, reasons to act, reasons why certain events happen (or not): in all cases, reasons are contrastive. A core motivation for Sinnott-Armstrong's contrastivism is that it can "illuminate examples . . . and . . . resolve or avoid puzzles and paradoxes" (135). Puzzles arise when questions about reasons are posed or claims about reasons are made without specifying any contrast class. If one does specify a contrast class, the puzzle disappears. In this chapter, Sinnott-Armstrong argues that "freedom" is a contrastive notion, along two dimensions. One should ask: "Free from what?", and one should ask "Free to do what?" As regards the first question, the idea is that if one is free, one is always free from certain constraints even if one is not free from other constraints. The advantages of this account are twofold. First, it is supported by common language. Second, as Sinnott-Armstrong argues, it can show that the argument from determinism commits the fallacy of equivocation. This account of freedom is Pyrrhonian in that the Pyrrhonian contrastivist suspends belief about which contrast class really is relevant. As regards the second question, here the idea is that one is free to do X as opposed to a contrast. One might be free to stop drinking whiskey in contrast with wine, but not free to drink only water in contrast with alcohol. Again, there is no such thing as "plain freedom." And again, Sinnott-Armstrong defends a Pyrrhonian account of this type of contrastive freedom. Sinnott-Armstrong ends this chapter by showing that Pyrrhonian contrastivism with respect to freedom comes in degrees and by considering an application of this brand of contrastivism.

Julia Driver, in her contribution "Luck and Fortune in Moral Evaluation" (Chapter 8), confronts the problem of moral luck in a contrastivist way. The problem of moral luck arises out of the following two claims. First, persons are only responsible for what they have control over. Second, we frequently don't have control over anything that happens as a result of, for instance, our actions. But now suppose that two persons perform exactly the same action, yet the consequences of the action of the one person turn out much worse than the consequences of the action of the other person. In such a case, the person whose actions turn out worse will get blamed more severely. Still, the fact that this person's actions turned out worse seems a matter of luck. The increased blame therefore seems paradoxical. One receives discredit for something one, intuitively, doesn't deserve. One account of the paradoxicality of moral luck is an epistemic account, where it is argued that we cannot tell if the two persons are equally blameworthy because we don't have access to their inner states. Put differently, this account presents an internalist solution to the problem of moral luck: the moral quality of one's actions is determined solely by factors internal to agency, such as one's motives or intentions. The plausibility of this account presents a problem for objective consequentialism, a position that holds that, in case of actions, the right action is the action that produces the best outcomes. The challenge, then, for the objective consequentialist is to account for moral luck without giving in to internalism. Driver's aim in her chapter is to meet this challenge while also clarifying what the problem of moral luck consists in and arriving at a better understanding of luck itself. Central to Driver's approach is that "luck" attributions are contrastive in nature: "S is lucky that p rather than q." But, Driver argues, a further variable should be introduced. Not only is luck contrastive; it is also relative to the agent's set of interests and epistemic states. Driver then goes on to discuss two approaches to luck: epistemic reductionism (luck reflects a state of ignorance on the part of either the luck attributer or the lucky individual) and the modal view (which holds that luck corresponds to "flukes"). Driver argues that both approaches should be taken contrastively and proposes that "we combine the intuitive appeal of the epistemic approach with the modal approach" (167). Finally, she shows how this can solve the problem of moral luck without going internalist.

The papers in this volume give an overview of the different types of contrastivism that can be defended. Probably, many more concepts could be contrastivized. But in order to assess the usefulness of doing so, I am confident that the papers in this volume will provide a good starting point.[2]

NOTES

1. As to contrastivism about knowledge, some key papers are Schaffer (2005b, 2007), Morton and Karjalainen (2003, 2008), Blaauw (2008a, 2008b, 2008c), Dretske (1970), and Sinnott-Armstrong (2008). A good overview is given by Morton (2010). Sinnott-Armstrong (2006) applies contrastivism to moral epistemology. There is contrastivism about causation. Some key papers are Schaffer (2005a, 2010), Maslen (2004), Woodward (2003), and Hitchcock (1996). As to contrastivism about explanation, key texts are Garfinkel (1981), Hitchcock (1996, 1999, 2001), Van Fraassen (1980), and Lipton (2004).
2. I am very grateful to Walter Sinnott-Armstrong for comments on this introduction.

REFERENCES

Blaauw, M. 2008a. Subject Sensitive Invariantism: In Memoriam. *The Philosophical Quarterly* 58(232): 318–326.

Blaauw, M. 2008b. Contesting Pyrrhonian Contrastivism. *The Philosophical Quarterly* 58(233): 471–478.

Blaauw, M. 2008c. Epistemic Value, Achievements, and Questions. *Proceedings of the Aristotelian Society (supplementary volume)* 82(1): 43–58.

Dretske, F. 1970. Epistemic Operators. *Journal of Philosophy* 67: 1007–1022.

Garfinkel, A. 1981. *Forms of Explanation: Rethinking the Questions of Social Theory*. New Haven: Yale University Press.

Hitchcock, C. 1996. The Role of Contrast in Causal and Explanatory Claims. *Synthese* 107: 395–419.

Hitchcock, C. 1999. Contrastive Explanation and the Demons of Determinism. *British Journal for the Philosophy of Science* 50: 585–612.

Hitchcock, C. 2001. The Intransitivity of Causation Revealed in Equations and Graphs. *Journal of Philosophy* 98: 273–299.

Karjalainen, A., and A. Morton. 2003. Contrastive Knowledge. *Philosophical Explorations* 6(2): 74–89.

Lipton, P. 2004. *Inference to the Best Explanation*. 2nd ed. London and New York: Routledge.

Maslen, C. 2004. Causes, Contrasts, and the Nontransitivity of Causation. In *Causation and Counterfactuals*, ed. J. Collins, N. Hall, and L.A. Paul, 341–357. Cambridge, MA: MIT Press.

Morton, A., and A. Karjalainen. 2008. Contrastivity and Indistinguishability. *Social Epistemology* 22(3): 271–280.

Morton, A. 2010. Contrastive Knowledge. In *The Routledge Companion to Epistemology*, ed. D. Pritchard and S. Bernecker. New York: Routledge.

Schaffer, J. 2005a. Contrastive Causation. *Philosophical Review* 114: 327–358.

Schaffer, J. 2005b. Contrastive Knowledge. In *Oxford Studies in Epistemology 1*, ed. T. Gendler and J. Hawthorne. Oxford: Oxford University Press.

Schaffer, J. 2007. Closure, Contrast, and Answer. *Philosophical Studies* 133(2): 233–255.

Schaffer, J. 2010. Contrastive Causation in the Law. *Legal Theory* 16: 259–297.

Sinnott-Armstrong, W. 2006. *Moral Skepticisms*. New York: Oxford University Press.

Sinnott-Armstrong, W. 2008. A Contrastivist Manifesto. *Social Epistemology* 22(3): 257–270.

Van Fraassen, B. 1980. *The Scientific Image*. Oxford: Clarendon Press.

Woodward, J. 2003. *Making Things Happen: A Theory of Causal Explanation*. Oxford: Oxford University Press.

1   Contrastive Explanation

Christopher Hitchcock

Numerous authors have claimed that the truth, or at least the felicity, of an explanatory claim is sensitive to contrast. For example, consider the following explanatory claims:

1. Adam ate the apple because he was hungry.[1]
2. Susan was arrested because she stole the bicycle.[2]

Fleshing out the details of the stories in appropriate ways, we might hear

1a. Adam ate the apple, rather than giving it back to Eve, because he was hungry

as true, and

1b. Adam ate the apple, rather than the pear, because he was hungry

as false. For instance, if Adam was sufficiently hungry that he would have indiscriminately eaten the first edible thing handed to him by Eve, we might accept 1a and reject 1b. Similarly, we might hear

2a. Susan was arrested because she stole the bicycle, rather than buying it

as true, and

2b. Susan was arrested because she stole the bicycle, rather than the skis

as false (where Susan robbed a sporting goods store). For instance, if the police were not singularly vigilant about catching bicycle thieves, and if Susan could not have used the skis to effect a rapid downhill getaway, we might accept 2a and reject 2b (but see section 3.1 below for a caveat).

Early discussions of this phenomenon include Hansson (1974), Dretske (1977), van Fraassen (1980), Garfinkel (1981), and Achinstein (1983). Some more recent attempts to explain the phenomenon include Lewis (1986a),

Hitchcock (1996, 1999), and Lipton (2004). In this paper, I will discuss the types of contrastive explanation and the sources of contrast-sensitivity in explanation. Of course, explanations frequently cite causal information, so the contrast-sensitivity of explanation is closely related to the contrast-sensitivity of causation, which is discussed in detail in Jonathan Schaffer's contribution to this volume (Schaffer 2012).

1   THE GRAMMAR OF EXPLANATION

Following Hempel (1965), I will say that an explanation is an answer to an explicit or implicit explanation-seeking "why?" question. Such a question has the canonical form:

Why is it the case that A?

where A is a proposition describing some event, state of affairs, or fact. A response to this question has the form:

A because B

where B is also a proposition or set of propositions. A is the *explanandum* and B is the *explanans*. I will argue in section 6 below that explanations in fact have a more complex structure, but this basic terminology will help to get us started.

Note that I do not assume that all "why?" questions are requests for explanations. Some are requests for reasons or justifications. If I say in exasperation, after my car breaks down on the freeway, "why does this always happen to me?", I would typically be understood as asking something like "what have I done to deserve this misfortune?", and not "what are the typical causal antecedents that lead to events like this one?" Likewise, "how?" questions can sometimes be requests for explanations, although they can also be requests for specifics about the nature of an event. If I ask, "how did George die?", I might be looking for an answer like: "He had severe arteriosclerosis. When he exerted himself by climbing the stairs, his heart could not receive enough oxygenated blood, and it failed." But I might (if I were feeling macabre) be looking for an answer like: "first he clutched his chest, then his eyes rolled into his head, then he keeled over backwards." Only the first of these answers constitutes an explanation of George's death.

Moreover, not all uses of the word "explanation" pick out the sorts of things that can be answers to explanation-seeking why questions. If I explain the rules of chess to you, I might be answering the question "how do you play chess?", but I am not answering the question "why do you play chess?" I will here be using the word "explanation" in the narrow sense, meaning the kind of thing that can be expressed by a proposition of the form "*A* because *B*" in response to an explanation-seeking "why?" question of the form "why is it the case that *A*?"

2 CONTRAST

There are a number of linguistic devices for inviting contrast. Perhaps the most simple is the phrase "rather than" as used in sentences 1a through 2b in the previous section. "Rather than" can be used to introduce contrast with one or more explicitly mentioned items, or with a class of items. Here are some examples:

3. Adam ate the apple rather than the pear.
4. Adam ate the apple rather than the pear or the banana.
5. Adam ate the apple, rather than one of the other fruits on the table.
6. Adam ate the apple rather than giving it back to Eve.
7. Adam ate the apple rather than doing something else with it.
8. Adam, rather than Eve, ate the apple.
9. Adam, rather than one of the other guests at the party, ate the apple.

Each of these sentences does three things: (i) it asserts that Adam ate the apple; (ii) it explicitly asserts, or at least directly implies, that various alternatives to Adam's eating the apple did not occur; and (iii) it invites a contrast between Adam's eating the apple and the other alternatives mentioned. Sentences 3 through 9 agree with respect to what they assert in (i), but they deny different alternatives in (ii) and invite different contrasts in (iii). For example, sentence 3 contrasts Adam's eating the apple with his eating the pear, whereas 4 contrasts it with his eating the pear and with his eating the banana. I will call the proposition asserted by the sentence the "focus," and the contrasting alternatives "foils."[3] Note that I said in part (ii) that each sentence *explicitly* asserts, or *directly* implies, that various alternatives do not occur. If our background theory entails that the various alternatives are incompatible—for example, if it implies that Adam will eat only one thing, or that only one person can eat the apple—then the non-contrastive sentence "Adam ate the apple" will *imply* that these alternatives did not occur, as will all of the sentences 3 through 9.

Note that "rather than" takes a noun phrase, rather than a clause, as its completion. We don't say:

3a. *Adam ate the apple rather than Adam ate the pear.[4]

This grammatical feature of our use of "rather than" is misleading. The contrast invoked is between two propositions or states of affairs. For example, sentence 3 invites a contrast between

Adam ate the apple

and

Adam ate the pear.

Sentence 3 does not assert that Adam ate some contrastive object, the apple-rather-than-the-pear.

There are a number of other expressions that function in essentially the same way as "rather than"—for example, "as opposed to" and "in contrast with." Thus the following sentences are synonymous with 3:

3c. Adam ate the apple, as opposed to the pear.
3d. Adam ate the apple, in contrast with the pear.

There are other linguistic devices for inviting contrasts. One is stress or emphasis. In written language, stress is indicated using <u>underlining</u>, **bold-face**, CAPS, or *italics* (which I will use from here on). In spoken language, it is indicated using increased volume and, in English at least, rising intonation. Here are some examples:

10. Adam ate the *apple*.
11. Adam *ate* the apple.
12. *Adam* ate the apple.

Sentences bearing stress invite contrasts with sentences that result from substituting alternatives for the stressed item. Thus 10 invites contrast with "Adam ate the pear," "Adam ate the banana," and so on; 12 invites contrast with "Eve ate the apple," "Abel ate the apple," and so on. Because these sentences don't mention explicit alternatives, the range of possible substitutions must be determined by context. For instance, context will determine whether Adam's eating the apple is to be contrasted with his eating one of the other fruits in the fruit bowl, or with his eating meat, cheese, bread, etc. Cleft and pseudocleft constructions function in much the same way:

13. It was the apple that Adam ate.
14. What Adam ate was the apple.
15. It was Adam who ate the apple.
16. The person who ate the apple was Adam.

An interesting and important linguistic question is whether contrast is a *semantic* or a *pragmatic* phenomenon. Is it part of the *meaning* of sentences 3, 10, and 13 that Adam's eating the apple is to be contrasted with his eating the pear, or is this contrast rather a matter of conversational *implicature*? As we will see in the next section, contrast is closely connected with presupposition. Linguists are still divided over whether presupposition is a semantic or pragmatic phenomenon. Fortunately, my discussion of the role of contrast in explanatory claims can remain neutral on this question.

3 PRESUPPOSITION

A *presupposition* of a sentence is a proposition that must be true in order for the sentence to make sense, or be appropriate. In conversation, presuppositions are propositions that are accepted, at least for the sake of argument, by all participants in the discussion, and which provide a background for the discussion. Consider, for example, the sentence

17. Stephen stopped smoking.

This sentence has as a presupposition

17P. Stephen used to smoke.

It might be natural to think that 17 *asserts* 17P; for instance, we might analyze 17 as

17'. Stephen used to smoke, but he no longer smokes.

The problem with this way of analyzing 17 is that it mischaracterizes the way 17P behaves when 17 is embedded. Consider the simple case of negation:

17N. Stephen has not stopped smoking.

If 17 is understood as 17', then 17N should be equivalent to

17N'. Either Stephen smokes now, or he didn't use to smoke.

In fact, however, 17N tends rather to imply that 17P is true. That is, 17P is entailed by both 17 and its negation. This is a characteristic feature of presuppositions: they are inherited in contexts where the original sentence is embedded.[5]

Consider now our sentences

3. Adam ate the apple rather than the pear;
10. Adam ate the *apple*.

3 has as a presupposition that Adam ate the apple or the pear, whereas 10 has as a presupposition that Adam ate something. Thus

3N.   Adam didn't eat the apple, rather than the pear

strongly suggests that Adam ate the pear, whereas

10N.   Adam didn't eat the *apple*

seems to imply that Adam ate something else. In general, a sentence that invites contrast tends to bear a presupposition that either the focus, or one of the foils, is true.

4   TYPES OF CONTRASTIVE EXPLANATION

Contrast in an explanatory claim can take on many different forms. I want to distinguish here between three different dimensions of variation: (i) wide versus narrow scope; (ii) contrast in the explanans versus contrast in the explanandum; and (iii) contrasting processes versus contrasting outcomes.

4.1   Wide versus. Narrow Scope

Consider the following explanatory claim, bearing contrastive stress:

18. Susan was arrested because she stole the *bicycle*.

If we ignore the stress, our explanandum is "Susan was arrested," and our explanans is "Susan stole the bicycle." The stress on "bicycle" invites contrast with statements in which "bicycle" is replaced with the name of some other item that Susan might have stolen. But there is an ambiguity. Does the stress invite alternatives to the entire sentence, or just alternatives to the explanans? If the former, we will say that the contrast has *wide* scope; if the latter, *narrow* scope. If the contrast has narrow scope, then the contrast takes place within the "because" sentential operator. If the contrast has wide scope, then 18 has a structure something like:

18w. (Susan was arrested because she stole the bicycle) contrasted with (Susan was arrested because she stole the skis) and so on.

However, if the contrast has narrow scope, the structure is more like:

18n. (Susan was arrested) because ([Susan stole the bicycle] contrasted with [Susan stole the skis] and so on).

There is an important difference in meaning, or at least in felicity conditions, between the two. 18w suggests that although Susan's arrest might, hypothetically, have been explained by her theft of the skis (or some other item), it was, in fact, the bicycle that she stole and which led to her arrest. In the following dialogue, 18 is being asserted in the sense of 18w:

"I heard that Susan was arrested because she stole the skis."
"No. Susan was arrested because she stole the *bicycle*."

By contrast, 18n suggests that the difference between stealing the skis (or some other item) and stealing the bicycle is explanatorily relevant to Susan's arrest. One way of drawing this distinction out is to notice that 18w and 18n differ in their counterfactual import. 18w suggests that Susan would have been arrested had she stolen the skis (or some other item) instead of the bicycle. Perhaps it would be more accurate to say that 18w is most felicitously uttered when this counterfactual is believed or supposed to be true. By contrast, 18n suggests that Susan would *not* have been arrested had she stolen the skis (or some other item).

I do not think that contrast with wide scope introduces any serious new problems for the theory of explanation. If we have an adequate account of explanation when no contrast is involved, then we can analyze 18w along something like the following lines:

  (i)   if Susan had stolen the bicycle, then (a) she would have been arrested, and (b) her theft of the bicycle would explain her arrest;
 (ii)   if Susan had stolen the skis, then (a) she would have been arrested, and (b) her theft of the skis would explain her arrest; (and so on for the other contrastive foils);
(iii)   in fact, Susan stole the bicycle.

There may be some fine tuning: We may want to replace (ii) with the condition that this counterfactual is being supposed, or believed. Moreover, (ii), or whatever we replace it with, is much more plausibly viewed as an assertion condition than a truth condition. But the key point is that an account of non-contrastive explanation can be imported into the analysis of 18w without modification.

On the other hand, I think that the sensitivity of explanatory claims to contrast with narrow scope tells us something deep about the nature of explanation itself. All contrastive claims will be understood as having narrow scope in what follows.

4.2   Explanans versus Explanandum

Consider two of the claims with which we began:

 1a. Adam ate the apple, rather than giving it back to Eve, because he was hungry.
 2a. Susan was arrested because she stole the bicycle, rather than buying it.

In 1a, the contrast occurs in the *explanandum*. That is, the explanandum, Adam's eating the apple, is contrasted with an alternative, Adam's giving it back to Eve. In 2a, the contrast occurs in the *explanans*: the explanans, Susan's stealing the bicycle, is contrasted with an alternative.

Most of the literature on contrastive explanation has focused on the case where there is contrast in the explanandum. For example, Hansson (1974), van Fraassen (1980), Lewis (1986a), and Lipton (2004) discuss only this case. On the other hand, discussion of cases like 2a first entered the literature in discussions of causation, rather than explanation (e.g. in Dretske 1977). This seems odd, because the two cases seem closely related. Indeed, one can easily construct explanatory claims with contrast in both positions:

> 19. Susan was arrested, rather than let off with a warning, because she stole the bicycle, rather than the tennis ball.

So it seems that an account of contrastive explanation that can only handle one of these types of contrast must be inadequate. The account that I will develop below can account for contrast in the explanandum, explanans, or both.

4.3   Process versus Outcome

There are two different kinds of case where there is contrast in the explanandum. Compare our recurring example

> 1a. Adam ate the apple, rather than giving it back to Eve, because he was hungry

with the following example:[6]

> 20. Smith, rather than Jones, developed paresis, because Smith had latent, untreated syphilis.

Some background: The example of paresis entered the literature on explanation in Scriven (1959). "Paresis" here presumably refers to what is more usually called "general paresis" or "general paresis of the insane"; in normal medical usage, "paresis" *simpliciter* refers to partial paralysis that can have a variety of causes. For simplicity, however, I will stick with the less accurate "paresis." Paresis is a neurological disorder typically manifesting in psychotic symptoms, caused by advanced syphilis infection. The point of the example, for Scriven, was that although only a minority of those infected with syphilis develop paresis, only those with syphilis develop paresis. The point that van Fraassen makes is that we would accept 20 if Smith had syphilis but Jones did not, but we would not accept it if both men had syphilis.

Comparing 1a and 20, both contrast what actually happened with an alternative that did not happen. 1a contrasts Adam's eating the apple with an alternative in which he instead hands it back to Eve. 20 contrasts Smith's developing paresis with Jones's developing paresis. But the nature of the contrasting alternatives is different. In 1a, the contrast involves the very same process—Eve handing the apple to Adam, his considering it, and so on—leading to a different outcome. In 20, the contrast is with a different process—Jones instead of Smith, with all of the various ways in which Jones is different from Smith—leading to the same outcome, paresis. For this reason, I will describe the type of contrast in 1a as a *same process/ different outcome* contrast, and that of 20 as a *different process/same outcome* contrast. (Reminder: the contrast is always with an alternative that did *not* occur. In 1a the same process did *not* yield a different outcome. In 20, the different process did *not* yield the same outcome.)

One difference between the two types of contrast is that in same process/ different outcome contrasts, the alternatives are incompatible, or at any rate, competing. Perhaps it was not strictly impossible for Adam to both eat the apple and give it back to Eve—perhaps he could have eaten half of it, and given it back, or perhaps he could have given it back, then eaten it out of her hand—but the strong suggestion of 1a is that only one of the two outcomes was going to occur. In 20, however, there is no incompatibility between Smith developing paresis and Jones developing paresis. Smith's paresis does not somehow confer immunity upon Jones, or draw the paresis away from him.

It may not always be obvious whether a contrast in the explanandum is a same process/different outcome contrast, or a different process/same outcome contrast. For example, if Smith and Jones were competing for the affections of the same woman, who (unbeknownst to them) was the only eligible woman in the community who had syphilis, 20 might plausibly be construed as involving a same process/different outcome contrast. That is, an explanation of why 20 is true might be an explanation of why the woman chose Smith, rather than Jones.

5 A PICTURE OF EXPLANATION

I will now present a picture of explanation. I say a "picture" rather than a theory, because the account I will give will be abstract, and short on details. I think that the details can be filled out in a number of different ways—in fact, they will have to be filled out in different ways for different kinds of explanation. Thus the picture is compatible with many different specific theories of explanation.

An explanation provides us with information about what the explanandum *depends on*. It gives us information about how changes in the conditions described in the explanans would result in differences in the explanandum.

To use a phrase of Woodward's,[7] an explanation provides resources for answering "what if things had been different?" questions. I will illustrate this idea using a number of different types of explanation.

5.1   Ordinary Causal Explanations

Ordinary causal explanations proceed by citing one or more causes of the explanandum. Causes are typically events on whose occurrence the explanandum counterfactually depends. Of course Lewis (1973) attempted to turn this idea into an analysis of causation. Similarly, Woodward's (2003) interventionist account of causal explanation relies on certain kinds of counterfactuals. There are problems that remain, particularly involving cases of causal preemption, but we will relegate them to the realm of "details." Even if is not possible to analyze causation in terms of counterfactuals, it remains the case that citing causes typically does yield information about counterfactuals. Thus if I say that Adam ate the apple because he was hungry, I would naturally be understood as saying that Adam's eating the apple *depended* upon his hunger, in the sense that he would not have eaten it had he not been hungry. Perhaps the relevant counterfactuals are more subtle: perhaps the level of Adam's hunger determined how much of the apple he ate, or perhaps his eating the apple depended upon his hunger only when certain other events are held fixed.[8] The bare explanatory claim does not really discriminate between these possibilities. In general, ordinary causal explanations are fairly blunt instruments for conveying information about how the explanandum might have been different.

5.2   Quantitative Causal Explanations

Here is an example that Woodward uses on a number of occasions (e.g. 2003). A positively charged particle is located at a certain distance from a long wire with uniform negative charge. It accelerates toward the wire (Figure 1.1). Using Coulomb's law, we can calculate the force acting on the particle, and hence the acceleration. But the law does not merely tell us that whenever we have a particle with this charge, at this distance from a wire with this charge distribution, it will accelerate in this way. It also allows us to calculate how the particle would accelerate if it were at a different distance, if the charge distribution were different, or even if the wire were shaped differently. All of these elements—the position of the particle, the charge and shape of the wire—might naturally be regarded as causes of the particle's acceleration. But the explanation does not merely tell us that if one of these elements had been different, the particle would not have accelerated in the same way. It provides us with the resources to determine exactly what the acceleration would have been had the circumstances been different.

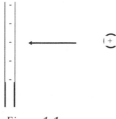

Figure 1.1

5.3   Probabilistic Explanations

Suppose that a vertically polarized photon interacts with a polarizer that is oriented at an angle of θ from the vertical. The photon passes through the polarizer. We can explain this using the Born rule from quantum mechanics. This tells us that the photon has a probability of $\cos^2\theta$ of passing through the polarizer. Although this explanation cannot tell us what definitely would have happened if the angle of displacement between the photon and the polarizer had been θ' instead (unless θ' is a multiple of $\pi/2$), it does tell us how the probability of transmission depends upon the angle of displacement.

5.4   Functional Explanations

Why do elephants have such large ears? Elephants are large mammals that live in hot climates, so it is important for them to have a method of dissipating heat. Blood vessels running through the ears allow the blood to be exposed to the air for rapid cooling. This is a *functional* explanation: it explains why elephants have large ears by citing their function. Following Wright (1976), it has become common to understand functional explanation as a kind of causal explanation. The elephant's large ears cause heat to dissipate, and the dissipative capacity of the large ears has caused them to be selected by natural selection. Like other causal explanations, functional explanations provide us with information about conditions under which the explanandum would have been different. For example, this functional explanation tells us that elephants would have smaller ears if they evolved in a cooler climate.

5.5   Constitutive Explanations

Sometimes we explain the macroscopic properties of an object, organism, or substance by describing its internal structure. For example, we might explain why diamonds are so hard in terms of the configuration of the carbon atoms that make up diamonds. Each carbon atom bears a strong covalent bond with four other carbon atoms. The four other atoms are equidistant, located at the points of a tetrahedron (Figure 1.2). This symmetric structure makes the network of carbon atoms highly resistant to strain in all directions. This is not a causal explanation in the ordinary sense. We usually require that causes be genuinely distinct from their effects (see e.g. Lewis 1986b). Here the micro-structure of the diamond is not really distinct from its hardness. Rather, the microstructure *constitutes* the hardness of the diamond. Nonetheless, the explanation does provide us with information about conditions under which the explanandum would be different. If the structure of the lattice of atoms is not symmetric, so that the bonds are weaker in some directions than in others, the substance will not be resistant to certain types of strain. For instance, in graphite, the carbon atoms are arranged in "sheets." Each atom has strong covalent bonds with three other atoms that are co-planar with it, but only weak bonds with atoms that lie off the plane (Figure 1.3). This means that graphite offers little resistance to strains that are parallel to the sheets, allowing them to slip past one another.

There is an interesting problem of how exactly to understand the dependence of the diamond's hardness on its microstructure. For example, if we are working within Woodward's interventionist framework, it does not seem that we can independently intervene on the microstructure and the macro-properties of diamond. An intervention on the molecular structure *just is* an intervention on the diamond's hardness, and vice versa. So there are important problems about how, precisely, to implement this account of constitutive explanation. But the general framework seems to me to be right.

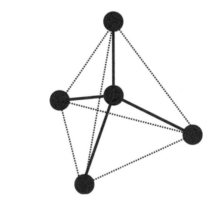

Figure 1.2

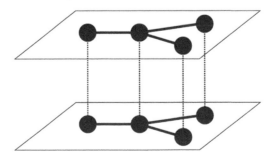

Figure 1.3

5.6 Theoretical Explanations

There is another kind of case where a scientific theory can describe a kind of dependence that cannot be easily construed as causal. For example, suppose we want to know why there are stable planetary orbits. It can be shown mathematically that stable orbits arise only in space-time structures of certain dimensionality. Four-dimensional space-times admit of stable orbits. The proof can be used to show what would happen if space-time had three dimensions or five, instead of four.[9] But it would be a stretch to call the dimensionality of space-time a *cause* of orbital stability. For example, it makes little sense to talk of intervening on the dimensionality of space-time. Nonetheless, there is a clear sense in which the stability of planetary orbits *depends* upon the dimension of space-time.[10] As with the case of constitutive explanation, there is a puzzle about where the asymmetry of explanation comes from. Whether or not planets have stable orbits (and the character of their orbits more generally) will mathematically entail facts about the dimensionality of space-time, but we would not say that stable orbits explain why space-time is four-dimensional. But I think that this is a problem for this type of theoretical explanation generally, not just with my take on it.

5.7 Mathematical Explanations

Sometimes we explain a phenomenon in terms of its abstract structure, which can be studied mathematically. Pincock (2007) gives a nice example. Why has no one succeeded in walking a continuous path through Königsberg, crossing each of the seven bridges exactly once? Königsberg is shown schematically in Figure 1.4. The answer can be understood if we think of Königsberg as instantiating the structure of a graph, where each land mass (both banks and both islands) is a node, and each bridge is an edge (Figure 1.5). In order to traverse a graph, traveling along each edge exactly once, the graph must be connected (it must be possible to get from every node to every other node, with the exception of nodes that

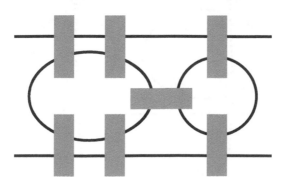

Figure 1.4

Figure 1.5

have no edges), and at most two of the nodes can have an odd number of edges connecting them. The reason is that any path on the graph has a starting node, and an ending node. All other nodes are intermediate nodes. Each time a path enters an intermediate node, it must also exit that node. If a path traverses each edge exactly once, then every intermediate node must be connected by an even number of edges. In the graph of Königsberg, however, all four nodes are connected by an odd number of edges.

Once again, the explanation tells us how the explanandum could be different. For instance, if we eliminate the bridge between the two islands (Figure 1.6), or add a bridge to each island (Figure 1.7), the conditions of the theorem will be met, and it will be possible to walk a continuous path that crosses each bridge exactly once.

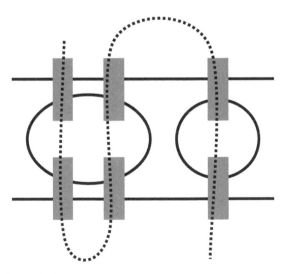

Figure 1.6

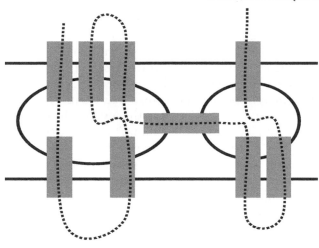

Figure 1.7

6   A REVISED GRAMMAR OF EXPLANATION

In light of this picture of explanation, I want to suggest a general schema
for explanation.[11]

ES: $X_1 = x_1, \ldots, X_n = x_n$
$\quad\ \ Y = f(X_1, \ldots, X_n)$
$\quad\ \ Y = y\ (\, = f(x_1, \ldots, x_n))$

$Y = y$ is the explanandum. $X_1 = x_1, \ldots, X_n = x_n$ describe particular facts or
conditions, and the equation $Y = f(X_1, \ldots, X_n)$ describes the way in which
Y depends upon X_1 through X_n; collectively these form the explanans. In
my schema, $X_1, \ldots, X_n$, Y are *variables*, not in the sense of the bound
and unbound variables of logic, but in the sense of the random variables
of measure theory. I will call $X_1, \ldots, X_n$ the *explanans* variables and Y
the *explanandum* variable. The variable Y, for example, might represent
the acceleration of a particle, with the values y corresponding to possible
values of the acceleration. In this case, Y is a quantitative variable, but a
variable may also be qualitative. For example, Y might represent the thing
that Adam ate, with values representing the apple, the pear, and so on. Formally, a random variable in measure theory is a function from the *outcome
space*, which we can think of as the set of all possible worlds, to the range
of the variable. I think of the variables in ES in much the same way. For
example, a variable AE representing what Adam eats maps worlds in which
Adam eats the apple to the value "apple," worlds where he eats the pear to
the value "pear," and so on. Note, however, that construed in this way, the
variables that figure in explanations will typically be *partial* functions. The

variable AE, for example, will only be defined on worlds where Adam eats something. We will call the proposition that a variable Y takes some value the *presupposition* of that variable. These presuppositions will function much like the linguistic presuppositions discussed in section 3 above.

It is possible for different variables to intersect. For example, let AE be a variable representing possible things that Adam eats, and let EA be a variable representing who might eat the apple. Thus AE will have as possible values "apple," "pear," "banana," etc., whereas EA will have as possible values "Adam," "Eve," "Abel," etc. Now both AE = "apple" and EA = "Adam" represent the same proposition, that Adam ate the apple. But they may function differently in an explanation. For example, EA may depend upon some other variable in a way that AE does not. This difference is also reflected in the fact that the two variables have different presuppositions.

The equation $Y = f(X_1, \ldots, X_n)$ represents the way in which the value of variable Y depends upon the value of $X_1, \ldots, X_n$. Representing this dependence with an equation is somewhat misleading, for mathematical equality is a symmetric relation, whereas the sorts of dependence relations that figure in explanations are typically asymmetric. The equation in ES is like the modifiable structural equations used in causal modeling (see e.g. Pearl 2000), where the asymmetry arises from restrictions on the ways in which values can be substituted for the variables in the equations. Indeed, when an explanation schema like ES represents a causal explanation, we may take the equation of the second line to *be* a structural equation of the appropriate sort. In particular, we may understand the equation $Y = f(X_1, \ldots, X_n)$ to entail that every normal, causal, non-backtracking counterfactual of the form

> If X_1 had been $x_1, \ldots,$ and X_n and x_n, then Y would have been $f(x_1, \ldots, x_n)$)

(where x_i is in the range of X_i for all i) is true. For theoretical explanations or constitutive explanations, however, the equation will have to represent a different sort of asymmetric dependence.

There are two ways in which ES can be generalized. First, in the case of probabilistic explanations, the equation of the second line can be replaced with one like the following:

> $Pr\,(Y) = f(X_1, \ldots, X_n)$

where $Pr\,(Y)$ is a probability distribution over the values of Y. Second, in may be possible to chain instances of ES into a narrative explanation, where the variables appearing above the line in one instance appear below the line in another instance. In the case of causal explanation, this chaining process results in a *causal model* of the sort described by Pearl (2000).

When communicating explanations verbally, however, one rarely provides an explicit schema like ES. Instead, one describes the explanans and explanandum using ordinary sentences, allowing context and verbal cues to provide information about the underlying variables. Moreover, the form of the dependence relation in the second line of ES is often presented in a merely qualitative and implicit way.

7   CONTRASTIVE EXPLANATION

We are now in a position to see why explanations are sensitive to contrast. Contrast plays the role of specifying which variables are involved in the explanation. (I leave it open whether this is achieved via the semantics or the pragmatics of contrastive statements.) Thus, for example, if we say

3. Adam ate the apple rather than the pear,

this tells us that the variable is AE, whose values range over possible things Adam eats, restricted to two possible values, the apple and the pear. 3 also tells us that AE takes the value "apple." Similarly,

10. Adam ate the *apple*

describes the variable AE', whose values also range over possible things Adam eats, but whose range is more inclusive that that of AE. 10 also tells us that AE' takes the value "apple."

An explanation will present one or more explanans variables and an explanandum variable, and will assert, minimally, that the value of the latter depends upon the values of the former. It might also tell us something about the form of the dependence. An explanation will be defective if it specifies explanans variables and an explanandum variable, and mischaracterizes the dependence between them. In particular, it will be defective if the explanandum variable does not depend upon the explanans variable(s) at all.

We will now see how this apparatus works with three different kinds of contrastive explanation: first, explanations where the contrast is in the explanans; second, explanations where contrast in the explanandum is between different outcomes of the same process; and finally, explanations where contrast in the explanandum is between different processes yielding the same outcome.

7.1   Contrast in the Explanans

Consider one of our examples from the introduction. Making plausible assumptions about the background story, we might hear

2a. Susan was arrested because she stole the bicycle, rather than buying it

as true, and

2b. Susan was arrested because she stole the bicycle, rather than the skis

as false.

Although neither 2a nor 2b has explicit contrast in the explanandum, it is reasonable to take the explanandum variable to be SA (Susan is arrested), taking possible values "yes" and "no." In 2a, the explanans variable is SB (what Susan did to the bicycle), taking possible values "steal" and "buy." 2a asserts, correctly, that the value of SA depends upon the value of SB (where the dependence is causal, or ordinary, non-backtracking counterfactual dependence). That is, 2a correctly asserts that if Susan had bought the bicycle, she would not have been arrested. In 2b, the explanans variable is SS (what Susan stole), taking possible values "the bicycle" and "the skis." 2b incorrectly implies that SA depends upon SS; hence it is false or defective.

7.2   Same Process/Different Outcome

Consider our other example from the introduction. With details fleshed out in a plausible way,

1a. Adam ate the apple, rather than giving it back to Eve, because he was hungry

sounds true, whereas

1b. Adam ate the apple, rather than the pear, because he was hungry

sounds false.

We may take the explanans variable to be AH (Adam is hungry), with possible values "yes" and "no." In 1a, the explanandum variable is AA (what Adam does to the apple), with possible values "eat" and "give back." 1a correctly tells us that the value of AA depends upon the value of AH. Specifically, it tells us that if AH had been "no," then AA would have been "give back." In 1b, the explanandum variable is AE (what Adam eats), with possible values "apple" and "pear." Now there is a sense in which the value of AE does depend upon the value of AH: namely that if AH had been "no," AE would not have taken any value at all. (That is, if Adam had not been hungry, he would have eaten neither the apple nor the pair.) But that is not the kind of dependence that an explanation is normally understood as describing. Rather, 1b asserts that *which value* AE takes depends upon AH.

Another way of stating this point is that if the dependence function f represents ordinary causal/counterfactual dependence, then in the equation

$$AE = f(AH),$$

the value of AE is undefined when AH = "no." This is analogous to the way in which the equation $Y^2 = X$ (where both X and Y are real-valued) leaves the value of Y undefined when X is negative. In general, an explanation will be defective when one or more values of the explanans variable(s) are incompatible (in whichever sense of incompatibility is appropriate for the particular explanation—for example causal or mathematical) with the presupposition of the explanandum variable.[12]

7.3 Different Process/Same Outcome

Finally, let us return to the following example from section 4:

20. Smith, rather than Jones, developed paresis, because Smith had latent, untreated syphilis.

Recall that the driving intuition behind this example is that the explanation is correct if Smith had latent, untreated syphilis, and Jones did not; however, 20 is defective if both men had latent, untreated syphilis.

Lipton (2004) has offered an influential account of contrastive explanation in terms of what he calls the "Difference Condition":

DC: To explain why P rather than Q, we must cite a causal difference between P and *not-Q*, consisting of a cause of P and the absence of a corresponding event in the history of *not-Q*. (Lipton 2004, 42)

Note, first, that this account only applies to contrast in the explanandum; it tells us nothing about contrast in the explanans. Second, the account is awkward, at best, in cases where the contrast involves the same process with a different outcome. Consider, for example,

1a. Adam ate the apple, rather than giving it back to Eve, because he was hungry,

which we have accepted as a successful explanation. According to DC, a successful explanation of why Adam ate the apple, rather than giving it back to Eve, must cite a cause of Adam's eating the apple, and the absence of a corresponding event in the history of his not giving it back to Eve. But what would be an event "corresponding" to Adam's hunger? If his hunger is an indiscriminate desire to eat, then perhaps a corresponding event would be an indiscriminate desire to give things to Eve.

Even if we grant this, does 1a really imply that Adam does not have such a desire? If 1a is intended to impugn Adam's generosity, it is certainly a cryptic complaint.[13]

Lipton's *DC* does, however, provide a fairly plausible account of contrastive explanation involving different processes leading to the same outcome. Applying it to 20, *DC* requires that latent, untreated syphilis be a cause of Smith's paresis, and that a corresponding event in the history of Jones's failure to develop paresis be absent. A "corresponding event" here is naturally understood as latent, untreated syphilis in Jones. Thus, the explanation is successful if Jones does not have latent, untreated syphilis, but not if he does.

The basic idea of Lipton's *DC* can be captured in my framework by interpreting the contrast as providing information about the range of one or more explanans variables. In 20, the explanandum variable is *SP* (Smith gets paresis), with values "yes" and "no." The explanans variable is *SS*, representing whether Smith has latent, untreated syphilis. But the *range* of this variable—the possible values that it can take—is restricted to the actual value, and the value of the corresponding variable possessed by Jones. Thus, if Jones does not have latent, untreated syphilis, *SS* can take the value "yes" or "no," and 20 correctly tells us that the value of *SP* depends upon the value of *SS*.[14] However, if Jones does have latent, untreated syphilis, then *SS* is the trivial variable that can only take the value "yes." In this case, trivially, the value of *SP* cannot depend upon the value of *SS*, and the explanation is defective.

Successful contrastive explanations of this kind pick out what Waters (2007) calls *actual difference-makers*. All causes are difference-makers in the sense that *hypothetical* differences in the cause would lead to *hypothetical* differences in the effect.[15] This is just what it is for an effect to depend counterfactually upon its cause. A cause is an actual difference-maker if actual differences in the values of a causal variable across different individuals in a population lead to actual differences in the values of effect variables. Some concepts that are naturally construed as causal are better understood in terms of actual difference-makers than in terms of causes *simpliciter*. For example, it might be natural to think that a trait is *heritable* if a parent's having that trait causes her offspring to have that trait. But in genetics, the quantitative notion of *heritability* is defined in terms of the actual variation in that trait. Thus a trait is heritable only to the extent that a parent's having that trait is an actual difference-maker for her offspring's having the trait. This difference is important, because only in the latter case can natural selection act on the trait.

8   CONTRASTIVE EXPLANATION AND DETERMINISM

Railton (1981), Salmon (1984), and Lewis (1986a) all claim that it is impossible to provide contrastive explanations of indeterministic events. More specifically, they seem to all claim that it is impossible to provide contrastive explanations involving different outcomes of the same process, when the outcome is chancy. The problem, they claim, is not indeterminism *per se*. Unlike Kitcher (1989) or Glymour (2007), for example, they have no problem with the explanation of indeterministic events in general. Lewis, for example, writes:

> We are right to explain chance events, yet we are right also to deny that we can ever explain why a chance process yields one outcome rather than another. (1986a, 230)

This strikes me as an odd tension in their accounts of explanation, and the present framework rejects this view almost completely.

Consider the example from section 5.3, where a photon interacts with a polarizer. The angle of polarization of the photon is displaced from the orientation of the polarizer by an angle of θ, resulting in a probability of $\cos^2\theta$ of passing through the polarizer. In fact it does pass through. Assume, with Railton, Salmon, and Lewis, that explanation is possible at all in this case. Let us consider some possible contrastive explanations. First, a case with contrast in the explanans:

21. The photon passed through the polarizer because it had a polarization oriented with angle θ to the polarizer, rather than angle θ'.

We may take the explanandum variable to be the result of the interaction of the photon with the polarizer (technically a *measurement* of the polarization of the photon), having possible values "transmitted" and "absorbed." The explanans variable is the displacement of the polarization of the photon to the orientation of the polarizer, restricted to the values θ and θ'. Because the probability of transmission does depend upon the angle, this is a perfectly acceptable explanation, so long as $\cos^2\theta'$ is different from (or perhaps less than) $\cos^2\theta$. If, for example, $\theta' = \pi/2$, 21 seems obviously correct.

Next consider a case where the contrast is between different outcomes of the same process.

22. The photon passed through the polarizer, rather than being absorbed, because it had a polarization oriented with angle θ to the polarizer.

Once again, the explanandum variable is the result of the interaction—transmission or absorption—and the explanans variable is the angle (this time unrestricted in range). Again, because the explanandum variable depends (probabilistically) upon the explanans variable in the appropriate way, the explanation is perfectly acceptable.[16]

Finally, consider a case where the contrast is between different processes yielding the same outcome.

23. Photon *a*, rather than photon *b*, passed through the polarizer, because photon *a* had a polarization oriented with angle θ to the polarizer.

If photon *b* has a polarization angle of θ', this yields essentially the same explanation as 21, and it will be acceptable so long as $\cos^2\theta'$ is different from (or perhaps less than) $\cos^2\theta$. Again, if photon *b* had a polarization of $\pi/2$, this explanation seems obviously correct. If photon *b* also had a polarization angle of θ, then 23 will be defective. But this has nothing to do with indeterminism *per se*. If there is no actual difference, there is no actual difference-maker. This applies to deterministic and indeterministic contexts alike.

There is, however, one grain of truth in the views of Railton, Salmon, and Lewis. Suppose that photons *a* and *b* are identical, and that photon *a* is transmitted, whereas photon *b* is absorbed. Now there can be no explanation of why *a* rather than *b* was transmitted. Because they are identical, there is no causal variable that can serve as an actual difference-maker. There is no analogous situation in deterministic contexts: if two different processes yield different outcomes, then there must be some causal difference between them. Note, however, that it is the different process/same outcome contrasts that pose the problem, not the same process/different outcome contrasts as Railton, Lewis, and Salmon claim. Moreover, it is only a very specific type of contrast, one where the processes are identical, that poses the problem. Apart from this very special case, there is no general problem with contrastive explanations of chancy outcomes.

NOTES

1. Based on an example in van Fraassen (1980).
2. Based on an example from Dretske (1977).
3. This terminology differs slightly from the more common "fact" and "foil" (as used e.g. by Lipton 2004). I avoid "fact," which has a number of other connotations. For instance, Bennett (1988) and Mellor (2004) argue that facts are the relata in causal relations.
4. I employ the standard convention from linguistics of using an asterisk to denote ungrammatical sentences.
5. See e.g. Soames (1989) for detailed discussion.
6. Discussed, e.g., by van Fraassen (1980).
7. See his (2003), as well as numerous earlier publications cited therein.
8. See e.g. Hitchcock (2001).
9. I ignore the possibility, envisioned by versions of string theory, that space-time might actually have ten or eleven dimensions.
10. This example was previously discussed in Hitchcock and Woodward (2003), where it was noted that this approach also fits well with Steiner's (1978) account of mathematical explanation.
11. This schema was suggested earlier in Woodward and Hitchcock (2003).
12. For further discussion of how the presupposition constrains successful explanation, see my (1996).
13. For further critique of Lipton's *DC*, see my (1999).

14. Recalling that not all of those with latent untreated syphilis develop paresis, the dependence function might rather tell us that the probability distribution on *SP* depends upon the value of *SS*.
15. Putting aside problems involving preemption.
16. For more detailed discussion of this case, see my (1999).

REFERENCES

Achinstein, P. 1983. *The Nature of Explanation*. Oxford University Press: Oxford.

Bennett, J. 1988. *Events and Their Names*. Indianapolis: Hackett.

Dretske, F. 1977. Referring to Events. *Midwest Studies in Philosophy* 2: 90–99.

Garfinkel, A. 1981. *Forms of Explanation: Rethinking the Questions of Social Theory*. New Haven: Yale University Press.

Glymour, B. 2007. In Defense of Explanatory Deductivism. In *Causation and Explanation*, ed. J.K. Campbell, M.O. Rourke, and H. Silverstein, 133–154. Cambridge, MA: MIT Press.

Hansson, B. 1974. Explanations—of What? Unpublished typescript.

Hempel, C.G. 1965. Aspects of Scientific Explanation. In *Aspects of Scientific Explanation and Other Essays in the Philosophy of Science*, 331–496. New York: Free Press.

Hitchcock, C. 1996. The Role of Contrast in Causal and Explanatory Claims. *Synthese* 107: 395–419.

Hitchcock, C. 1999. Contrastive Explanation and the Demons of Determinism. *British Journal for the Philosophy of Science* 50: 585–612.

Hitchcock, C. 2001. The Intransitivity of Causation Revealed in Equations and Graphs. *Journal of Philosophy* 98: 273–299.

Hitchcock, C., and J. Woodward. 2003. Explanatory Generalizations, Part II: Plumbing Explanatory Depth. *Noûs* 37: 181–199.

Kitcher, P. 1989. Explanatory Unification and the Causal Structure of the World. In *Scientific Explanation*, ed. P. Kitcher and W. Salmon, 410–505. Minneapolis: University of Minnesota Press.

Lewis, D. 1973. Causation. *Journal of Philosophy* 70: 556–567.

Lewis, D. 1986a. Causal Explanation. In Lewis 1986c, 214–240.

Lewis, D. 1986b. Events. In Lewis 1986c, 241–269.

Lewis, D. 1986c. *Philosophical Papers, Volume II*. Oxford: Oxford University Press.

Lipton, P. 2004. *Inference to the Best Explanation*. 2nd ed. London and New York: Routledge.

Mellor, D.H. 2004. For Facts as Causes and Effects. In *Causation and Counterfactuals*, ed. J. Collins, N. Hall, and L.A. Paul, 309–323. Cambridge, MA: MIT Press.

Pearl, J. 2000. *Causality: Models, Reasoning, and Inference*. Cambridge: Cambridge University Press.

Pincock, C. 2007. A Role for Mathematics in the Physical Sciences. *Noûs* 41: 253–275.

Railton, P. 1981. Probability, Explanation, and Information. *Synthese* 48: 233–256.

Salmon, W. 1984. *Scientific Explanation and the Causal Structure of the World*. Princeton: Princeton University Press.

Schaffer, J. 2012. Causal Contextualisms: Contrast, Default, and Model. This volume.

Scriven, M. 1959. Explanation and Prediction in Evolutionary Theory. *Science* 130: 477–482.

Soames, S. 1989. Presupposition. In *Handbook of Philosophical Logic*, vol. 4, ed. D. Gabbay and F. Guenthner, 553–616. Dordrecht: Reidel.

Steiner, M. 1978. Mathematical Explanation. *Philosophical Studies* 34: 135–151.

Van Fraassen, B. 1980. *The Scientific Image*. Oxford: Clarendon Press.

Waters, K. 2007. Causes That Make a Difference. *Journal of Philosophy* 104: 551–579.

Woodward, J. 2003. *Making Things Happen: A Theory of Causal Explanation*. Oxford: Oxford University Press.

Woodward, J., and C. Hitchcock. 2003. Explanatory Generalizations, Part I: A Counterfactual Account. *Noûs* 37: 1–24.

Wright, L. 1976. *Teleological Explanations*. Berkeley and Los Angeles: University of California Press.

2 Causal Contextualism

Jonathan Schaffer

> Causal statements are commonly made in some context, against a background which includes the assumption of some *causal field*. A causal statement will be the answer to a causal question, and the question 'What caused this explosion?' can be expanded into 'What made the difference between those times, or those cases, within a certain range, in which no such explosion occurred, and this case in which an explosion did occur?' Both causes and effects are seen as differences within a field. (Mackie 1974, 34–35)

Causal claims are context sensitive. For instance, if the engineer finds that the poor road conditions contributed to the accident, then it would be acceptable for her to say:

1. The poor road conditions caused the accident

Yet if the detective wants to focus on the drunk driver, then it would seem acceptable for him to deny 1 and instead say:

2. The poor road conditions didn't cause the accident, it was the drunk driver

So much is commonplace. As Lewis notes:

> We sometimes single out one among all the causes of some event and call it 'the' cause, as if there were no others. Or we single out a few as the 'causes', calling the rest mere 'causal factors' or 'causal conditions'. . . We may select the abnormal or extraordinary causes, or those under human control, or those we deem good or bad, or just those we want to talk about. (1986, 162)

Yet, despite extensive studies of context sensitivity for other aspects of language such as knowledge ascriptions, there has been little discussion of the context sensitivity of causal claims. I will address three questions. In section 1, I will address the question of whether the context sensitivity of causal claims is partly semantic, or wholly pragmatic. I will argue—in a way familiar from arguments for epistemic contextualism—that the context sensitivity of causal claims is partly semantic since it does not fully fit the pragmatic mold. In section 2, I will consider the question of whether causal claims are sensitive to contrasts, defaults, and/or models. I will argue that treating causal claims as sensitive to contrasts (for both cause and effect) does all the needed work. Finally in section 3, I will face the question—naturally arising from my answers to the first two questions—of how semantic sensitivity to contrasts might be implemented within an overall plausible semantic framework. This will turn out to be something of a puzzle. Accordingly, I must conclude that we do not yet have a clear understanding of context sensitivity as it arises for causal claims.

For those familiar with the discussion of the context sensitivity of knowledge ascriptions, it might be worth flagging two main respects in which the context sensitivity of causal claims will prove to differ. On the one hand, the intuitive data for context sensitivity is much stronger and more robust for causal claims, and includes specific phenomena that seem to have no counterpart with knowledge ascriptions (for instance, the matter of *selection* by which one causal factor is promoted to cause and the remainder demoted to background conditions). On the other hand, the semantic implementation of context sensitivity turns out to be far more problematic for causal claims, at least given the sort of contrastive views I advocate. This is because knowledge ascriptions only need a single source of contrasts which arguably can be read off the question under discussion. But causal claims need at least two separate sources of contrasts, and there is no obvious general procedure to recover the specific contrast applicable to the cause, or to recover the specific contrast applicable to the effect.

1   PARTLY SEMANTIC, OR WHOLLY PRAGMATIC?

Causal claims are context sensitive. That is, it may be acceptable for one speaker in one context to make a given causal claim, and acceptable for another speaker in another context to deny that very claim. This is uncontroversial. But what is controversial is whether such context sensitivity is a purely pragmatic affair, entirely explained by the extent to which the causal claim constitutes a cooperative contribution to the conversation at hand; or whether there is some semantic aspect to this context sensitivity. For instance, when the engineer finds that the poor road conditions contributed to the accident and says:

1. The poor road conditions caused the accident

And the detective denies 1 to lay the blame on the drunk driver, can it be that both the engineer and the detective still speak *truly*? Or must at least one of these characters (presumably the detective) be uttering a felicitous falsehood?

More precisely, what is at issue is the following thesis:

Causal Contextualism: A single causal claim can bear different truth values relative to different contexts, where this difference is traceable to the occurrence of 'causes,' and concerns a distinctively causal factor.

The first clause of *Causal Contextualism* characterizes the form of context sensitivity at issue: variation in truth value for a single sentence at multiple contexts of utterance. The remaining clauses try to ensure that this variation is arising for the right reason: not due to some other element of the sentence (perhaps all sentences contain other context sensitive elements), and not due to irrelevant features of 'cause' (such as tense and mood). This definition could perhaps use refinement, but should be clear enough to put to work.

1.1   The Context Sensitivity of Causal Claims

In order to evaluate *Causal Contextualism*, it will prove useful to provide a range of illustrations of the context sensitivity of causal claims. I do not claim that these illustrations exhaust all the context sensitivity of causal claims (I doubt they do), or that they must all receive a unified theoretical treatment (though I will offer one in terms of sensitivity to contrasts). Rather my purpose is to exhibit a family of striking and pervasive context sensitivities in causal discourse, in order to consider whether they fully fit the pragmatic mold. When I speak of "the context sensitivity of causal claims" in what follows, I should be understood as speaking of the sort (or sorts) of context sensitivity exhibited in these illustrations.

To begin, there is context sensitivity with respect to *causal selection*. It is part of causal discourse to promote some handful of factors to the status of cause, and to demote the remaining factors to the status of background condition.[1] This is the phenomenon seen in the case of the car accident above, and in Hart and Honoré's example of the Indian famine:

The cause of a great famine in India may be identified by an Indian peasant as the drought, but the World Food Authority may identify the Indian government's failure to build up food reserves as the cause and the drought as a mere condition. (1985, 35–36).

To provide one more illustration, the forest rangers would presumably promote the lightning strike to the status of cause for the forest fire, and would demote the presence of oxygen to the status of background condition. But now consider Putnam's visiting Venusians: "Imagine that Venusians land on earth and observe a forest fire. One of them says, '*I* know what caused that—the atmosphere of the darned planet is saturated with oxygen'" (1982, 150).

So in particular, we can imagine a conversation among the Venusians in which the following claim was acceptable:

3. The presence of oxygen caused there to be a forest fire

Yet if we imagine a conversation among the forest rangers, 3 will surely be unacceptable in such a context. The forest rangers will deny that the presence of oxygen caused the fire. In this vein Hart and Honoré note:

> In most cases where a fire has broken out the lawyer, the historian, and the plain man would refuse to say that the cause of the fire was the presence of oxygen, though no fire would have occurred without it: they would reserve the title of cause for something of the order of a short-circuit, the dropping of a lighted cigarette, or lightning. (1985, 11)

Further, there is context sensitivity with respect to *causal inquiry*. Causal claims are answers to 'why'-questions, and differences in the preceding 'why'-question may trigger differences in the acceptability of the resulting causal claims (van Fraassen 1980). For instance, if the question arises as to why John *kissed* Mary (perhaps we are wondering about John's courage in matters of love), then a causal answer should explain why John offered Mary a kiss rather than, say, a hug or a handshake. On the other hand, if the question arises as to why John kissed *Mary* (perhaps we are wondering about John's attraction to Mary), then a causal answer should explain why the person John kissed was Mary rather than, say, Suzy or Billy.

So in particular we can imagine a conversation in which we are presupposing that John loves Mary but questioning his romantic courage, in which the following claim might well be acceptable:

4. John's boldness caused him to kiss Mary

Yet if we imagine a conversation in which we are presupposing John's romantic courage, but questioning why he was attracted to Mary, an utterance of 4 may well be unacceptable. In such a context, one wants to hear about some feature of Mary (such as her sense of humor, or her flowing hair) that distinguishes her from her rivals.

Moreover, context sensitivity arises when there are *multiple alternatives* (Hitchcock 1996). For instance, suppose that the train switch has three settings. Setting it to *broken* will send the train down the broken track and on to disaster, setting it to *local* will send the train down the local track and slowly to the station, whereas setting it to *express* will send the train down the express track and swiftly to the station. The switch gets set to *local* and—just as expected—the passengers arrive slowly at the station. Did the switch's getting set to *local* cause the passengers to arrive at the station (not to arrive slowly, just to arrive at all)? The answer seems to be: *it depends on which other option you had in mind*.

So in particular we can imagine a conversation in which the background assumption is that the switch was set to *broken*, and we are wondering why disaster was averted. In such a context the following claim should be acceptable:

5. The switch's getting set to *local* caused the passengers to arrive at the station

After all, if we were expecting the train to derail, learning that the switch got set to *local* should help us understand why things went otherwise. Yet if we imagine a context in which the background assumption is that the switch should be set to *express* (and leaving it on *broken* is not under consideration), then an utterance of 5 should be unacceptable. After all, in such a context we were already expecting the passengers to arrive at the station. The switch's getting set to local makes no difference.

1.2   Sentential Sensitivities

So far I have illustrated three sorts of context sensitivity for causal claims. It will also prove useful to display some other "nearby" sensitivities in causal discourse which are not sensitivities of a single sentence to context, but rather sensitivities between different sentences employing distinct but still coreferential event descriptions. The issue in these cases is how the differences in the event descriptions impact acceptability, and the hope is that these cases might shed light on how contextual differences impact acceptability.

With this in mind, consider the role that explicit 'rather than' clauses can play in causal discourse. For instance, in the train case above, one wants to say that the switch's being set to *local* rather than *broken* caused the passengers to arrive at the station:

6. The switch's getting set to *local* rather than *broken* caused the passengers to arrive at the station

But equally one wants to deny that the switch's being set to *local* rather than *express* caused the passengers to arrive at the station (they would arrive safely either way), by denying:

7. The switch's getting set to *local* rather than *express* caused the passengers to arrive at the station

Yet unless one has an implausibly fine conception of events, it seems that the switch's getting set to *local* rather than *broken*, and the switch's getting set to *local* rather than *express*, pick out the same event under a different description. It is not as if the switch got set twice.

Moreover, it is not as if the switch's being set to *local* rather than *express* made no difference. Its being set to *local* caused the passengers to arrive at the station slowly rather than swiftly:

8. The switch's getting set to *local* rather than *express* caused the passengers to arrive at the station slowly rather than swiftly

And again, unless one has an implausibly fine conception of events, it seems that the passenger's arriving at the station, and the passenger's arriving at the station slowly rather than swiftly, pick out the same event. How are the 'rather than' clauses impacting acceptability, if the same events are picked out either way?

Or consider the role that more specific descriptions of events can play in causal discourse. To borrow an example from McDermott (1995), one might deny that McEnroe's tension caused him to serve, but accept that it caused him to serve awkwardly:

9. McEnroe's tension caused him to serve
10. McEnroe's tension caused him to serve awkwardly

That is, while 9 seems unacceptable, 10 is fine. One wants to say: the tension didn't matter to *whether* he served but only to *how*. Yet unless one has an implausibly fine conception of events, it seems that McEnroe's serving *just was* his serving awkwardly. We are just discussing a single serve.[2]

Indeed, as the following example from Achinstein (1975) shows, merely shifting the locus of focus within the event nominal can control acceptability:

11. Socrates's DRINKING HEMLOCK at dusk caused his death
12. Socrates's drinking hemlock AT DUSK caused his death

11 seems acceptable, but 12 does not. One wants to say: what Socrates drank mattered, when he drank it did not. Again, unless one has so fine a conception of events that focal differences can make for event differences, the same event of Socrates's drinking hemlock at dusk is described in both 11 and 12, merely with a difference in emphasis. There was just one drinking.

1.3   The Invariantist Orthodoxy

The question is whether the contextual and sentential sensitivities just illustrated indicate any sort of *semantic* context sensitivity in causal discourse (as per *Causal Contextualism*), or whether they can be fully explained via conversational pragmatics. Perhaps all one sees is differences in the extent to which a given causal claim is a cooperative contribution to different conversations.

To the extent that there is an orthodox view in the current literature, it is the view that the context sensitivity of causal claims is entirely pragmatic. This view denies *Causal Contextualism* without denying the sensitivity "data," instead positing a purely pragmatic explanation for this data:

> *Causal Invariantism*: It is not the case that a single causal claim can bear different truth values relative to different contexts, where this

difference is traceable to the occurrence of 'causes,' and concerns a distinctively causal factor. Causal claims are context sensitive in their acceptability, but the context sensitivity of causal claims is a wholly pragmatic phenomenon.

The first sentence of *Causal Invariantism* is the denial of *Causal Contextualism*, and the second sentence adds the acceptance of context sensitivity plus the posit of a purely pragmatic explanation for such context sensitivity.[3]

For the invariantist, causal discourse involves a preselective semantics for some egalitarian and unselective notion of *being a causal factor*. In the forest fire case above, both the lightning strike and the presence of oxygen should equally qualify. Indeed presumably all the positive causal claims made in sections 1.1–1.2, even if unacceptable in the context at hand, will count as true. In this vein, Lewis—while defending a counterfactual analysis of causation—clarifies that he is "concerned with the prior question of what it is to be one of the causes (unselectively speaking). My analysis is meant to capture a broad and nondiscriminatory concept of causation" (1986, 162).[4]

The invariantist than layers a selective pragmatics for *being a salient causal factor* (sometimes expressed as *being "the cause"*) over her preselective semantics. In the forest fire case, our interests and background expectations will determine which of the many "causes" gets selected as salient. Along these lines, Mackie speaks of causal selection as "reflecting not the meaning of causal statements, but rather their conversational point" (1974, 35), and Lewis explicitly associates causal selection with Gricean conversational pragmatics:

> There are ever so many reasons why it may be inappropriate to say something true. It might be irrelevant to the conversation, it might convey a false hint, it might be known already to all concerned, and so on (Grice 1975). (2004, 101; cf. Bennett 1995, 133)

Of course it is uncontroversial that there are pragmatic phenomena in discourse, and *a fortiori* uncontroversial that there are pragmatic phenomena in causal discourse. The question is whether pragmatics can fully explain the contextual and sentential sensitivities exhibited. To my knowledge no invariantist has ever tried to spell out the pragmatic explanations in any detail, or do much more than allude to the prospect of some Gricean story.

From a wider perspective, *Causal Contextualism* might be counted as orthodoxy, and rooted in Mill's groundbreaking discussion of causal selection. For Mill is a *revisionist* about causal discourse. He thinks that our causal claims are shot through with selection effects. He merely regrets this as unscientific and deserving of excision:

> Nothing can better show the absence of any scientific ground for the distinction between the cause of a phenomena and its conditions, than

the capricious manner in which we select from among the conditions that which we choose to denominate the cause. (1950, 244)

So when it comes to a purely descriptive account of our causal concept, Mill looks to be on the contextualist side. Contextualism also has roots in Hart and Honoré's discussion of the role of causation in the law: "The contrast of cause with mere conditions is an inseparable feature of all causal thinking, and constitutes as much the meaning of causal expressions as the implicit reference to generalizations does" (1985, 12). Contextualism has further roots in van Fraassen's (1980) discussion of the context sensitivity of explanation, insofar as both causal and explanatory claims are understood as triggered by contrastive why questions. And contextualism seems to have attracted a fairly wide range of contemporary theorists, including Hitchcock (1996), Woodward (2003), Maslen (2004), Menzies (2004 and 2007), Schaffer (2005a and 2010), Hall (2007), and Northcott (2008). So perhaps a new orthodoxy is (re-)forming.

1.4   Against Invariantism

Evidently there are pragmatic phenomena in discourse, and *a fortiori* there are pragmatic phenomena in causal discourse. The question is whether pragmatics serves to fully explain the context sensitivity of causal claims. I will now offer three connected arguments that conversational pragmatics cannot fully explain this context sensitivity. (Given that *Causal Invariantism* is the main alternative to *Causal Contextualism*, these arguments are indirectly arguments for the contextualist alternative.)

The first argument is that *no known pragmatic mechanism handles the cases*. So suppose that there is a lightning strike and a forest fire breaks out, and that a forest ranger utters 3, citing the presence of oxygen as causing the fire. Given that this is unacceptable, and given that the pragmatic explanations available are going to involve something like Gricean maxims—and in particular floutings of Gricean maxims which produce false implicatures—one can ask which Gricean maxim is flouted.[5] The only Gricean maxim which seems applicable is Relevance. Indeed it seems clear that the remaining Gricean maxims—namely Quality, Quantity, and Manner—need not be flouted at all. After all, the forest ranger might have excellent evidence for there being oxygen present and for it being a factor, she has been informative, and she has not spoken in an overly prolix or otherwise marked manner.

So I take it that the only known pragmatic mechanism that might be operative in this case is Relevance, and that the invariantist will say that when the forest ranger is inquiring into the causes of the forest fire, she is *presupposing* that oxygen is present, and wondering about *what ignited the oxygen*. So citing the presence of oxygen is failing to speak to the question under discussion, and hence flouts Relevance.[6]

But this pragmatic explanation—which looks like the only one available in anything like a Gricean framework—is inadequate, because floutings of Relevance produce a distinctive feel not found in the illustrations, which is a not a feeling of falsity but merely of irrelevance. Thus consider Kierkegaard's (1978, 50) parable of the madman who repeats "Bang! The earth is round!" at every turn. One is inclined to label the man mad and his utterances irrelevant, *but there is no feeling that the man has said anything false.* We clearly recognize what the madman keeps repeating as an irrelevant *truth.* (We also recognize that prefacing every utterance with "Bang!" is a bit odd, but leave that aside.) No one, on considering Kierkegaard's madman, should feel any inclination to reject the claim that the earth is round.

Matters may be clearest in the examples with sentential differences from section 1.2. For instance, on the pragmatic view 7 and 9 are literally true but merely irrelevant:

7. The switch's getting set to *local* rather than *express* caused the passengers to arrive at the station
9. McEnroe's tension caused him to serve

Yet neither seems like an irrelevant truth which simply did not bear mentioning. Instead both feel more like falsehoods. The switch's getting set to *local* rather than *express* did not cause the passengers to arrive at the station—it made no difference whatsoever to whether the passengers arrived at the station, because both settings are stipulated to result in this same outcome. Likewise McEnroe's tension did not cause him to serve—it made no difference whatsoever as to whether he served, because he was set to serve anyway. The distinctive feel of irrelevant truth is absent.

The second argument against pragmatic explanations is that *speakers assert the negations.* Ordinary speakers will not only refuse to assert claims like 3 (in the context of the forest rangers); they will go so far as to assert its denial:

13. The presence of oxygen did not cause there to be a forest fire, what caused the fire was the lightning

(The more sophisticated speaker may then clarify that the presence of the oxygen was a "mere background condition" or something of that ilk.) Likewise in the case of the car accident, the detective who wants to focus on the drunk driver will *deny* that the poor road conditions caused the accident, as per 2. Floutings of Relevance will at best explain a refusal to assert 1 or 3. They will not explain a willingness to assert the negation as seen in 2 or 11, since (by the lights of invariantists) the negations are equally irrelevant and false to boot![7]

Indeed the first two arguments against pragmatic explanations connect. When the pragmatic explanation involves a mere flouting of Relevance, the assertion will have the feel of an irrelevant truth. This is why there will be no temptation to assert the negation (an irrelevant falsehood). Consider again Kierkegaard's madman. No sane and minimally informed speaker would go so far as to assert the negation: "The earth is not round."

Third and finally, *cancellation does not help*. The main test for conversational implicatures is that they (unlike semantic entailments) are cancelable. For instance, if I say of a job candidate that she has excellent handwriting, I can block the implicature that she is a poor philosopher by saying "but I don't mean to suggest that she is a poor philosopher" (I may then go on to discuss her philosophical genius). None of the causal cases pass the test. Thus consider, in the context of the forest rangers:

14. The presence of oxygen caused there to be a forest fire, but I don't mean to suggest that the lightning strike played no role

This hardly seems any more acceptable, despite the cancelation of any potential implicature that the lightning strike played no role. Or consider:

15. McEnroe's tension caused him to serve, although I don't mean to suggest that it was the only factor involved

These attempts at cancelation hardly seem to salvage acceptability. With 15, one wants to say that regardless of which factors did cause McEnroe to serve, his tension was not among them.

These three arguments thus constitute a *prima facie* case against *Causal Invariantism* and thereby a *prima facie* case for its main competitor, *Causal Contextualism*. I do not mean to suggest that these arguments are decisive. The invariantist can always challenge the "data" or try to introduce new pragmatic mechanisms to better explain it, and there are also alternatives to consider between *Causal Invariantism* and *Causal Contextualism* (for instance, perhaps some of these judgments should be explained away as performance errors of some sort). But I do mean to suggest that the invariantist orthodoxy, which assimilates the context sensitivity of causal claims to Gricean implicatures, is implausible. As soon as one tries to spell out the details of the Gricean story, it emerges that the context sensitivity of causal claims does not fit the pragmatic mold. Perhaps Mill was right from the start in his descriptive claim about our causal concept.

2   WHAT SHIFTS?

So far I have argued that the context sensitivity of causal claims is partly semantic, or at the least is not wholly a matter of Gricean conversational pragmatics. But leaving this aside, there is a largely independent question of which contextual parameters causal claims are sensitive to. What shifts with context? That is, what gears of the contextual machinery are engaged by the illustrative cases above, on which the acceptability of the relevant causal claim turns?

To illustrate the sort of question I am asking, consider a simple indexical like 'I.' No serious account of 'I' could rest with the claim that it exhibited context sensitivity, or even with the claim that it exhibited semantic context sensitivity. There is the further question of which contextual parameters 'I' is sensitive to. In this case the answer is straightforward: the semantic value (content) of 'I' is sensitive to the contextual factor of *who is speaking*. (The reason why it is acceptable for me but not you to say 'I am Jonathan Schaffer' is that when I say it the 'I' refers to me and so the claim is true, but when you say it the 'I' refers to you and so the claim is false.) Presumably the context sensitivity of 'cause' is a sensitivity to some other factor or factors, but which?

Such a question is largely independent of the previous question as to whether the sensitivity is pragmatic or semantic, but not completely independent. For if 'cause' is sensitive to a given factor, then there must be a parameter at the relevant pragmatic or semantic level tracking this factor. With 'I,' given that it is sensitive to who is speaking, and given that this is a semantic matter, then there must be a semantic level parameter tracking who is speaking. With 'cause,' I will be arguing that it is sensitive to contrast, and given that this is a semantic matter (as previously argued), then there must be a semantic level parameter tracking the contrasts. I will return to this matter in section 3.

2.1   Contrasts, Defaults, and Models

What are causal claims sensitive to? It turns out that there are at least three different—albeit compatible and not wholly distinct—sorts of answers that one finds in the literature. One answer, which I will be defending (and which is defended in various forms in Hitchcock 1996; Woodward 2003; Maslen 2004; Schaffer 2005a; and Northcott 2008), is that causal claims are sensitive to *contrasts*. What shifts with context are the contrasts in play, where contrasts are specific possible alternatives to actual events. Actually there are at least four versions of contrastivism that are found in the literature, concerning whether one is looking at a contrast for the cause (c^*), for the effect (e^*), for both, or for each event in the set of events under consideration (V^*):

> *Cause-Contrast*: c rather than c^* causes e
> *Effect-Contrast*: c causes e rather than e^*
> *Double-Contrasts*: c rather than c^* causes e rather than e^*
> *Total-Contrasts*: c causes e relative to V^*

(One might also work with a set C^* of contrasts for the cause and/or a set E^* of contrasts for the effect, but I will suppress this complication for simplicity.)

The contrastive view can—though it need not—be plugged into a simple counterfactual test for causation by replacing the supposition of the nonoccurrence of *c* or *e* (or of any intermediaries or other events involved in the account), with the supposition of the occurrence of the associated contrast. So for instance—at least as a decent gloss of *Double-Contrasts*—one might hold that *c* rather than c^* causes *e* rather than e^* iff (roughly) if c^* had occurred then e^* would have occurred. I will be defending *Double-Contrasts* (though I would be equally happy with *Total-Contrasts*—what is crucial is just that we have contrasts for both cause and effect; further contrasts might also prove useful). The counterfactual test just offered will prove useful insofar as it—together with certain assumptions about which contrasts are relevant in which contexts—will allow one to use *Double-Contrasts* to test truth values for causal claims.

But a different answer (supported by Menzies 2004; Hitchcock 2007; and Hall 2007) is that causal claims are sensitive to *defaults*. What shifts with context are which outcomes count as the "normal" or "default" behavior of the system under consideration, and which count as "abnormal" or "deviant" behavior. It is theoretically possible to assign defaults to a range of possible outcomes for the cause, for the effect, for both, or for every event under consideration (just as with contrasts), but all the defaultists in the literature have worked with the idea that defaults are assigned for all events under consideration:

> *Default*: *c* causes *e* relative to *Def*.

Def is a function from each event under consideration to a range of "default" outcomes associated with that event (the actual event might be a default outcome or a deviant outcome).

A guiding idea behind *Default* is that causes and effects are conceptualized as deviations from some sort of natural state (Maudlin 2004). This idea can, for instance, be plugged into a simple counterfactual test by treating the nonoccurrence suppositions as reversions to default behavior. So for instance, where *Def* assigns a single default outcome for both *c* and *e*, one might hold that *c* causes *e* iff if *Def(c)* had occurred then *Def(e)* would have occurred.[8]

And yet a third answer (found in Menzies 2004; Halpern and Pearl 2005; Hitchcock 2007) is that causation is relative to an eligible causal model of the situation:

> *Model*: *c* causes *e* relative to *Mod*.

Mod may be a set of variables and structural equations as in Pearl (2000), or a set of objects assumed to form a closed system plus a set of governing laws as in Menzies (2004). This is the natural reading of any theorist in the causal modeling tradition who gives an account of when one variable in a model is causally related to another variable in a model, while allowing (as is usually allowed) that there are worldly situations for which there are multiple eligible causal models with diverging causal verdicts.

The various contrastive proposals, *Default*, and *Model* are not wholly distinct, and indeed—at least on their leading implementations—can be ordered in strength as follows:

Cause-Contrast Effect-Contrast	Double-Contrasts	Total-Contrasts	Model	Default

Working backwards, *Default*—at least as implemented in Hall (2007) and Hitchcock (2007)—is a strict addition to *Model*, since default structure is given by adding the *Def* function to Pearl models, augmenting the variables and structural equations of Pearl models with a function from each variable to some subset of its allotted values that are to count as its default settings. So implemented, default relativity might be understood as relativity to *augmented models* with an added *Def* function. And *Model* is an addition to *Total-Contrasts*, insofar as models include variables with a range of allotted values, which range is a contrast space for the event modeled by the variable.[9] *Model* adds a further relativity to other aspects of causal models beyond the range of allotted values for variables, namely the choice of events modeled by variables and the structural equations over the variables. *Total-Contrasts* adds to *Double-Contrasts* a further relativity to contrasts for events under consideration other than cause or effect, and *Double-Contrasts* adds to both *Cause-Contrast* and *Effect-Contrast* a relativity to contrasts for the other side of the causal relation.

This means that *Total-Contrasts* can be thought of as partial model relativity. *Total-Contrasts* can be thought of as relativity to the range of allotted values for the variables, *without* relativity to the remaining aspects of the model, namely which events are represented by variables, and what structural equations hold over these variables. With respect to the events represented, it is natural to think that there is an objective fact as to which events are out in the world to be represented. Models which—for the sake of tractability—do not represent all the events idealize at their peril. With respect to the structural equations, these are generally supposed to hold objectively, representing the counterfactual facts as to what would lead to what. Fix which variables are modeled and what range of values they are allotted, and there is a right choice of structural equations. Any relativity to "a different choice" of structural equations is at best mistaken. Perhaps *Total-Contrasts* thus captures the element of truth in *Model*, while avoiding the other implausible aspects of model relativity. Though I will largely work with *Double-Contrasts* in what follows, my openness to the strengthened thesis of *Total-Contrasts* largely stems from this connection to causal models.

Note also that contrastivity, default-relativity, and model-relativity are compatible, and so one might endorse any combination thereof, including:

Contrast-Default-Model: c rather than c^* causes e rather than e^* relative to *Def* and *Mod*

Though—at least on the leading implementations of these ideas, on which they are ordered in strength as per above—such combinations are not genuinely new options. Given that default relativity includes model relativity (and thereby includes contrast relativity), *Contrast-Default-Model* is just *Default* by another name.

Note further that these options are hardly exhaustive. They are merely the main options that have been considered in the literature. One could also endorse an ambiguity thesis on which 'cause' can express a plurality of these candidates (Hitchcock 2003). That said, I will be arguing that contrastivity—and specifically *Double-Contrasts*—suffices to explain the context sensitivity of causal claims, so there seems no need (at least with respect to the cases currently under discussion) for anything further or stronger, or for any posited ambiguity. (Though again I am officially neutral between *Double-Contrasts* and *Total-Contrasts*.)[10]

I will further argue (section 2.3) that contrast sensitivity is specially rooted in the theoretical roles that causation plays. And I will argue (section 3) that there is independent linguistic reason to think that contrasts are elements of conversational context, and so are available as a contextual parameter to connect with causal claims. This claim does not carry over to defaults or models (or other arbitrary proposals). So the contrastive view also seems uniquely well situated with respect to linguistic implementation. Alas, I will also be arguing that the contrastive view faces linguistic difficulties as well, so I must be wary of claiming any ultimate advantage on this last matter.

2.2   Context Sensitivity as Contrast Sensitivity

I will now argue that contrastivity—and specifically *Double-Contrasts* on which causation is a relation of the form c rather than c^* causes e rather than e^*—serves to explain the context sensitivity of causation (section 1.1), and the nearby sentential sensitivity (section 1.2). Recall that the context sensitivity of causation, at least in the form under discussion, encompasses:

- causal selection (as illustrated by whether or not the presence of oxygen is said to cause there to be a forest fire)
- causal inquiry (as illustrated by the different causal answers appropriate for the questions of why *John* kissed Mary, why John *kissed* Mary, and why John kissed *Mary*)
- multiple alternatives (as illustrated by the train switch with the *broken*, *local*, and *express* settings)

And the nearby sentential sensitivity, at least in the form under discussion, encompasses:

- 'rather than'-clauses (as illustrated by 'the train switch being set to *local* rather than *express*' as opposed to 'the train switch being set to *local* rather than *broken*')

- event specifications (as illustrated by 'McEnroe's serving' versus 'McEnroe's serving awkwardly')
- focus shifts (as illustrated by comparing 'Socrates DRINKING HEMLOCK at dusk' with 'Socrates's drinking hemlock AT DUSK')

I will now argue that these sensitivities are all connected manifestations of an underlying contrast sensitivity in causal discourse.[11]

It might help to start with the focus shift cases, since these are perhaps clearest in terms of the theoretical treatment required. Focus (at least of the sort exhibited in the cases at hand) is *contrastive focus*. Returning to 11 and 12:

11. Socrates's DRINKING HEMLOCK at dusk caused his death
12. Socrates's drinking hemlock AT DUSK caused his death

In 11, 'Socrates's DRINKING HEMLOCK at dusk' is naturally interpreted as c: Socrates's drinking hemlock at dusk, rather than c^*: Socrates's drinking wine at dusk (or some other salient alternative to drinking hemlock); whereas in 12, 'Socrates's drinking hemlock AT DUSK' is naturally interpreted as c: Socrates's drinking hemlock at dusk, rather than c^*: Socrates's drinking hemlock at dawn (or some other salient alternative to occurring at dusk). Indeed such a contrastive treatment falls out of leading linguistic treatments of focus such as Rooth's (1992) alternative semantics, on which the focus semantic value of an expression is the result of replacing the focused constituent with the set of contextually salient options.[12] The difference between 11 and 12 is *not* between the actual events denoted, but between the contrasts selected.

Strictly speaking 11 and 12 only call for contrasts to the cause, as per *Cause-Contrast*. But it is easy to see that the same pattern can be found on the "effect side" as well, as seen in:

16. Socrates's DRINKING HEMLOCK at dusk at dusk caused HIS DEATH at dawn
17. Socrates's DRINKING HEMLOCK at dusk caused his death AT DAWN

16 seems acceptable but 17 does not, and these differ only in the locus of focus on the effect side. One wants to say that Socrates's drinking hemlock rather than wine (as per the contrastive interpretation on the cause side) made a difference to whether or not he died, but not to when he died—had Socrates drank wine he would have survived through the relevant time. (Though if the context is an unusual one in which the alternative of Socrates dying at a ripe old age is salient, then 17 should become acceptable. This is further confirmation of the way in which the contextual salience of contrasts controls acceptability.)

The shifts in 'rather than'-clauses—which are just overt contrastives—clearly follow the same pattern. Indeed 11, 12, 16, and 17 can all be rephrased—preserving the patterns of acceptability—with 'rather than'-clauses concerning the focused alternatives instead of focus. Or to return to the train cases, recall 8 (which has 'rather than'-clauses for both cause and effect):

8. The switch's getting set to *local* rather than *express* caused the passengers to arrive at the station slowly rather than swiftly

This is acceptable since the difference between *local* and *express* is what made the difference between a slow and swift arrival. But vary either of the 'rather than'-clauses to lose difference making and the result is unacceptable:

18. The switch's getting set to *local* rather than *broken* caused the passengers to arrive at the station slowly rather than swiftly
19. The switch's getting set to *local* rather than *express* caused the passengers to arrive at the station slowly rather than suffer a derailing

After all, with 18 the passengers were not going to arrive swiftly whether the switch was set to *local* or *broken* (it is not as if setting the switch to *broken* would have sped up their arrival!) And with 19 the passengers were not going to suffer a derailing whether the switch was set to *local* or *express* (either way they are safe). Moreover, the 'rather than'-clauses can still be re-correlated to re-gain difference making, with acceptability regained, as in:

20. The switch's getting set to *local* rather than *broken* caused the passengers to arrive at the station slowly rather than suffer a derailing

Given that the 'rather than'-clauses are overtly specifying the relevant contrasts (either directly providing the value of c^* and e^*, or—perhaps better—describing c and e in ways that naturally generate values for c^* and e^*), this is further direct evidence for an underlying contrast sensitivity in causal discourse.

Shifting to the event specificational differences, the very same pattern emerges. In the case of McEnroe's serve, the underlying contrastive causal truths (made explicit via overt 'rather than'-clauses) is as follows:

21. McEnroe's being tense rather than calm caused his serving awkwardly rather than gracefully

While the following is a contrastive causal falsehood:

22. McEnroe's being tense rather than calm caused his serving rather than standing still

All that needs to be added is that describing the effect event as a "serving"—as in 9—invites a contrast such as a standing still, and so invites an interpretation via the falsehood of 22. But describing the effect event more specifically as a "serving awkwardly"—as in 10—invites a contrast such as a serving gracefully, and so invites an interpretation via the truth of 21. Thus the difference in acceptability between 9 and 10 is naturally explained on a contrastive treatment.

So far I have argued that the sentential sensitivity of causal claims (section 1.2) is due to an underlying contrast sensitivity. It remains to show that the contextual sensitivity of causal claims (section 1.1) evinces the same underlying pattern. It might help to start on this point with the multiple alternatives seen in the train case, by reconsidering:

5. The switch's getting set to *local* caused the passengers to arrive at the station

The "data" observed in section 1.1 was that the acceptability of 5 seemed to vary with which alternative (*express* or *broken*) was salient. Assuming that the contextual alternative to arriving at the station is derailing, then on the contrastive treatment 5 is equivalent to the false contrastive claim 19 in contexts in which *express* is the salient contrast to the cause, and equivalent to the true contrastive claim 20 in contexts in which *broken* is the salient contrast.

Or consider sensitivity to the causal inquiry, as it impacts the acceptability of:

4. John's boldness caused him to kiss Mary

Note that different causal inquiries are associated with different contrastive why-questions. These generate contrasts on *e*, and generate different slates of permitted answers, which generate contrasts on *c*. Think of the contrasts as the contextually permitted answers to the twofold question 'What happened, and why?', where the 'What happened?' provides the space of salient options for the effect and the 'why?' provides the space of salient options for the cause.

So if we are questioning John's romantic courage the 'What happened?' aspect of the causal inquiry might present the options of (i) John kissed Mary, and (ii) John merely waved goodnight, and the 'why?' aspect of the causal inquiry might present the options of (iii) John is bold, and (iv) John is timid. In such a context 4 will be equivalent to the following contrastive truth:

23. John's being bold rather than timid caused him to kiss Mary rather than merely waving goodnight

Yet if we are not questioning John's romantic courage but instead questioning why he chose to kiss Mary, the 'what happened?' aspect of the causal inquiry might present the options of (i) John kissed Mary, and (ii*) John kissed Billy. In such a context 4 will be equivalent to the following contrastive falsehood:

24. John's being bold rather than timid caused him to kiss Mary rather than Billy

Or at least, 24 is false given that John's preference for Mary over Billy is not a matter of boldness. (If John is a confirmed homosexual who is boldly experimenting with his sexuality, then 4 should become acceptable. This is further confirmation of the way in which the contextual salience of contrasts controls acceptability.)

Finally, returning to causal selection, recall how this impacts the acceptability of claims such as:

3. The presence of oxygen caused there to be a forest fire

The question is *why* the Venusians naturally promote the presence of oxygen to the status of a cause while the forest rangers naturally demote the presence of oxygen to the status of a mere background condition. A natural first thought is that, for the Venusians, there is a salient alternative to the presence of oxygen: *the absence of oxygen*. But for the forest rangers no alternative to the presence of oxygen is salient. For the forest rangers the presence of oxygen is simply presupposed.

So understood, there is a single contrastive truth in play:

25. The presence rather than absence of oxygen caused there to be a forest fire rather than no fire

The reason why 3 is acceptable in the context of the Venusians is because it is equivalent to 25, since the absence of oxygen is a relevant alternative for them. But since the forest rangers recognize no salient alternative to the presence of oxygen, 3 does not receive any such interpretation (nor is it obvious what if any interpretation it should receive).[13]

On this treatment, causal selection stems from different background expectations which generate different causal inquiries. The forest rangers are presupposing that oxygen is present, and in effect asking what ignited the oxygen. The information about the presence of oxygen does not answer their question. The Venusians on the other hand are presupposing that lightning strikes are present, and in effect asking what the lightning strikes ignited. The information about the presence of oxygen answers their question. Overall what seems to be governing selection is the causal inquiry and its attendant possible answers (the contrasts). In causal selection what varies in a "capricious manner" (as Mill says) is which contrasts are in play in a given context, but what is predictable is what counts as the cause given the contrasts.

The above account of causal selection is essentially Mackie's view, on which a "causal statement will be the answer to a causal question" (1974, 34), and on which "causes and effects are seen as differences within a field" (1974, 35). The elements of the field are the contextually determined background conditions. Indeed, as I have suggested elsewhere (Schaffer 2005a), Mackie's view is the only plausible account of selection in the literature. If so then selection requires use of contrasts.[14]

Putting all of this together, *Double-Contrasts* seems capable of explaining all the contextual and sentential sensitivities of causation under discussion, and doing so in a unified and elegant way. Thus I would conclude that what shifts with context are the salient contrasts to the cause and to the effect.

2.3 Theoretical Motivations for Contrast Sensitivity

I have just argued—as an inference to the best explanation for the context sensitivity of causal claims—that causation is a contrastive relation, of the form c rather than c^* causes e rather than e^*. This conclusion may be buttressed by considering the theoretical roles of causation as a relation of *difference making*, as connected to *agential manipulation*, and as supporting *explanation*. Any relation that plays these roles needs contrastive structure.

As to difference making, recall what Lewis says in connecting counterfactuals to causal reasoning: "We think of a cause as something that makes a difference" (1986, 160–161). Lewis goes on to think of a cause as something whose occurrence or nonoccurrence makes a difference to the occurrence or nonoccurrence of the effect. But seen this way it should be evident that the notion of difference making is a contrastive notion. The contrasts articulate what the salient differences are. And it should be evident that the Lewisian notion of difference making involving nonoccurrence is just one of many ways of making a difference. There can also be a difference between a cause and an alternative to it (other than nonoccurrence), with respect to the effect versus an alternative to it (other than nonoccurrence).

Moreover, there is reason to think that the very idea of a "nonoccurrence" which Lewis appeals to is itself implicitly contrastive, in the sense that nonoccurrence suppositions take us *to the contextually salient alternative* (Schaffer 2005a). In this vein consider 'If *John* had not kissed Mary. . .'—one naturally imagines someone else doing the kissing. But consider 'If John had not *kissed* Mary. . .'—one naturally imagines something like a chaste handshake; or instead consider 'If John had not kissed *Mary*. . .'—now one naturally imagines John kissing someone else. Thus the expressions 'if *c* had not occurred' and also 'then *e* would not have occurred' in Lewis's counterfactual account are naturally read as equivalent to 'if *c** had occurred' and 'then *e** would have occurred,' where *c** and *e** are the contextually salient contrasts to *c* and *e* respectively.

As to agential manipulation, everyone accepts that there are connections between causation and the notions of intervention, manipulation, and agency. Never mind in which directions the connections run—all that matters here is that these notions are interconnected. Now the notion of manipulation seems patently contrastive, as Woodward explains:

> Any manipulation of a cause will involve a change from one state to some specific alternative, and how, if at all, a putative effect is changed under this manipulation will depend on the alternative state to which the cause is changed. Thus, if causal claims are to convey information about what will happen under hypothetical manipulations, they must convey the information that one or more specific changes in the cause will change the effect (or the probability of the effect). This in turn means that all causal claims must be interpretable as having a contrastive structure. (Woodward 2003, 146)

So it seems that causation must embody some sort of sensitivity to alternative courses of action ("hypothetical manipulations") if it is to properly connect to agency.

Finally, causation is widely thought to back explanation, and explanation has itself been argued to be contrastive (van Fraassen 1980; Garfinkel 1981). For instance, the explanation for why John kissed Mary rather than merely waving goodnight to her might differ from the explanation for why John kissed Mary rather than Billy. Or to express the matter with focus: the explanation for why John *kissed* Mary might differ from the explanation for why John kissed *Mary*. Given that causation serves to back explanation, it is most natural to posit that causal relations have the same contrastive structure as the explanations they serve to back.

The idea that causal claims are contrast sensitive is thus not *ad hoc* but rooted in the roles that the notion of causation plays. I should note that these role arguments might be thought to push from *Double-Contrasts* to *Total-Contrasts*, if for instance we are looking at cases involving difference-making *chains* where we need to think of the connection from *c* to *e* as mediated via *d*. But since I am maintaining neutrality between *Double-Contrasts* and *Total-Contrasts*, this is a matter I will leave for further discussion.

3   SEMANTICS FOR CONTRASTIVISTS?

So far I have argued that causal claims are semantically context sensitive as per *Causal Contextualism* (section 1), and that the sensitivity involved is sensitivity to the salient contrasts to c and e as per *Double-Contrasts* or *Total-Contrasts* (section 2). This picture of semantic sensitivity to contrasts invites a natural follow-up question, as to whether and how semantic sensitivity to contrasts can be implemented within an overall plausible semantic framework. This will turn out to be something of a puzzle, with two connected aspects.

The first aspect of the puzzle concerns the existence of any semantic level parameter or parameters that tracks the kind of *bi-contrastivity* I have posited for causal claims. While there is good reason to posit a semantic level parameter (namely *the question under discussion*) which generally provides for contrasts, it is much more difficult to motivate any general provision for two separate reservoirs of contrasts (*contrasts specifically for the cause* and *contrasts specifically for the effect*). The second aspect of the puzzle concerns the connection between the semantic level parameter or parameters and the denotation of 'causes.' Even given a general provision for two separate reservoirs of contrasts, the clause spelling out the denotation of 'know' must pick up on these parameters in a precedented and plausible way.

I remain hopefully that this twofold puzzle can be solved, but cannot yet offer anything like a satisfactory solution. Accordingly I must conclude that we do not yet have a clear understanding of context sensitivity as it arises for causal claims. This is everyone's problem. It arises in a specific form given the sort of semantic bi-contrastivity I have argued for. But the problem re-arises in different forms for different approaches. (For instance, if one thinks that there is merely pragmatic default sensitivity instead, one needs to show how this fits into an overall plausible pragmatic framework.) In this respect the context sensitivity of causal claims might be especially interesting to the student of context sensitivity, insofar as the data seems strong but the theoretical treatment difficult.

3.1 The Problem of Bi-Contrastivity

To begin with, there is good reason to posit a semantic parameter which generally provides for contrasts. This parameter is the *question under discussion*, posited as an element of the contextual scoreboard. The question under discussion (or perhaps better: a stack of questions, with the topmost element being under discussion) is widely posited to explain various phenomena such as topic choice and the licensing of ellipsis (Roberts 2004).

Questions are sets of alternatives. For instance, on the influential account of Groenendijk and Stokhof (1984), questions are *partitions* on logical space. So for instance, given that John, Billy, and Mary are the contextually salient individuals, the intensions of the question *who ate the cookies* is the set whose eight members are:

John	Billy	Mary	
Y	Y	Y	(John, Billy, and Mary all ate the cookies)
Y	Y	N	(John and Billy ate the cookies, but Mary did not)
Y	N	Y	
Y	N	N	
N	Y	Y	
N	Y	N	
N	N	Y	(Only Mary ate the cookies)
N	N	N	(No one ate the cookies)

Here is a space of contrasts, as a semantic parameter on the contextual scorecard. So it seems that semantic contrastivity is linguistically plausible (in a way that attributing default and model elements is not, pending any independently attested evidence that arbitrary contexts track such information).

Moreover, the idea that the causal contrasts are coming from the question under discussion directly fits the idea that the causal inquiry controls acceptability (seen in the case where John kissed Mary, and in Mackie's account of causal selection). If the causal inquiry forms the question under discussion and thus provides contrasts, and causation is a relation involving contrasts, then there is a direct link between the causal inquiry and the contrasts involved in causation.

But there are two problems with relying on the question under discussion to furnish the contrasts. The first and most glaring problem is that—at least on the form of contrastivism I have defended—one needs more than a single source of contrasts. One needs two separate sources of contrasts, one for the contrast to the cause and another for the contrast to the effect. One needs semantic bi-contrastivity, and this goes beyond what any one question can be guaranteed to provide.

The second problem is that it is not obvious that the question under discussion will provide any contrasts *for the cause* or *for the effect*. For instance, imagine that the king has eaten soup and died, and that the question under discussion is whether there is any connection between these events. In such a context, the occurrence of both the candidate cause and effect events are presupposed. It seems as if alternatives are not being queried as to what the king ate, or what fate he suffered. The alternative being queried are of the wrong sort entirely to provide either of the contrasts needed. (This shows that the problems are not solved by trying to retreat from *Double-Contrasts* to *Cause-Contrast* or *Effect-Contrast*.)

Full disclosure: I do not know how to solve these problems. But here is one thought which may not be hopeless, which involves thinking of causal claims as obligatorily triggering a specific sort of question under discussion. There are conjunctive questions which not only provide the kind of bi-contrastivity required, but which moreover specifically target the cause and effect slots. Most generally, such questions take the form:

26. What happened, and why?

The 'What happened?' aspect of 26 is understood to provide alternatives to the effect, and the 'Why?' aspect is understood to provide alternatives to the cause. For instance, in the case of the king just above, the true answer to "What happened?" is that the king died, and the other possible answers might include the option that the king is merely sleeping; while the true answer to "Why?" might be that the king ate the soup, and the other possible answers might be that the king poured the soup down the drain.

If it could be maintained that causal claims obligatorily trigger a question under discussion of the sort exemplified by 26, then all would be well (at least with respect to the problem of bi-contrastivity). But I should like some independent reason to maintain that causal claims obligatorily trigger such a question under discussion, beyond the fact that it would help me out. Any such triggered question ought to *show up* in topic choice and the licensing of ellipsis, in just the way that the question under discussion generally makes itself manifest in discourse (which provide the very rationale for positing the question under discussion parameter in the first place). I am not convinced that the question I am suggesting may be triggered shows up in the right ways.

3.2 The Problem of Connection

Suppose that the problem of bi-contrastivity (section 3.1) is somehow surmounted; there still remains a problem of how to connect the presence of contrasts in the context to the truth-conditions of the causal claim. After all, there are many elements of the contextual scorecard, and not every denotation is sensitive to every element. For instance, assuming that there is an element of the contextual scorecard for who is being addressed, presumably the denotation of 'and' remains contextually invariant, and the denotation of 'I' remains contextually variant but still invariant with respect to that parameter. So what is it about 'know' that connects it to that parameter, and thus enables it to pick up on the contextually given contrasts?

It has been argued that there are tight constraints between context sensitivity and logical form. In particular Stanley argues that all semantic context sensitivity arises from either indexicality or something like a covert variable: "Any contextual effect on truth-conditions that is not traceable to an indexical, pronoun, or demonstrative in the narrow sense must be traceable to a structural position occupied by a variable" (2000, 401). This is an attractive picture insofar as it provides principled constraints on context sensitivity, especially so given that there are principled tests for indexicals and for covert variables. So—assuming Stanley's constraints on semantic context sensitivity—there are three main options: either 'cause' is an indexical, or it projects covert contrast variables which may be evaluated by context, or it is not really semantically context sensitive after all.

I think it should be fairly clear that 'cause' is no indexical. Indeed it seem to fail all standard tests for indexicality.[15] For instance, we automatically adjust indexicals in indirect quotation. If Ann says 'I'm thirsty' we report 'Ann said that she is thirsty,' shifting automatically from her 'I' to our 'she.' We do *not* report homophonically by 'Ann said that I'm thirsty.' Nothing like this seems to occur with causal claims. If the engineer concerned with the roads says:

1. The poor road conditions caused the accident

Then it seems that she may be homophonically reported in any context, even the context of the detective concerned with the drunk driver, via:

27. The engineer said that the poor road conditions caused the accident

Or at least there is nothing like the smooth and automatic adjustment of indexicals across contexts.

So can the context sensitivity of causal claims be understood in terms of a covert variable (or perhaps two covert contrast variables) instead? Perhaps so, but again I should like some independent reason to posit such variables, beyond the fact that it would help me out. Any such variables ought to *show up* in standard tests for covert variables. But—at least with respect to *the binding test* (Partee 1989; Stanley 2000)—no such variables seem to turn up:

28. *Rather than any other setting, the switch's being set to *local* caused the passengers to arrive at the station.
29. *Rather than any other outcome, the switch's being set to *local* caused the passengers to arrive at the station.

Perhaps there are ways to explain binding failures compatible with the presence of the covert variable. Perhaps there are other syntactic diagnostics that would render a different verdict. But *prima facie* there does not look to be a covert variable in the syntax. And so it seems that—at least if Stanley's constraints are accepted—then there are good arguments against locating the context sensitivity of causal claims anywhere in the semantic machinery.

Since I am characterizing a parameter as semantic when it impacts truth-conditions, there remain several other options. One option would be to add contrasts parameters to the index by which propositions are evaluated for truth, alongside the orthodox world and time parameters posited by Kaplan. But this strikes me as unpromising, since in my view it was a mistake all along to have such an index at all (Schaffer *forthcoming*). At any rate the standard reason for wanting parameters—namely the existence of a sentential operator said to work by shifting them—does not seem to apply. Another option would be to allow for *free enrichment* whereby considerations of general rationality can add constituents that lack representation in logical form (Sperber and Wilson 1986). But the worry with such an option—which is a main motivation for Stanley's claim that "all truth-conditional effects of extra-linguistic context can be traced to logical form" (2000,. 391)—is that it overgenerates, undoing needed constraints.

3.3 Concluding Puzzlement

What emerges is an inconsistent triad of seemingly plausible claims:

> *Pragmatic or Semantic*: Causal claims are either pragmatically context sensitive or semantically context sensitive (section 1.1).
> *Not Pragmatic*: Causal claims are not pragmatically context sensitive (section 1.3).
> *Not Semantic*: Causal claims are not semantically context sensitive (sections 3.1–3.2).

Something must go. One either needs to reconsider all of the examples of section 1.1 so as to deny *Pragmatic or Semantic*; or one needs a better account of the pragmatic mechanisms in play, that will enable one to deny *Not Pragmatic*; or one needs a better account of how contextual elements can impact the semantics, that will enable one to deny *Not Semantic*. In other words, the context sensitivity of causal discourse seems to fit neither the Gricean view of pragmatics nor Stanley's constraints on semantic context sensitivity.

I hold out hope that *Not Semantic* can be answered. Or at least, it seems to me that the case for *Pragmatic or Semantic* is extremely strong, turning on "data" that has been universally accepted since Mill. And it seems to me that the case for *Not Pragmatic* is fairly strong as well, at least on anything like a Gricean picture. In contrast I think that the case for *Not Semantic* is a good deal weaker, involving controversial matters concerning the question under discussion and strong views on how truth conditions are constrained by syntax. But that said, these remain serious problems that I do not know how to resolve.

So I must conclude that there is as of yet no decent account of the context sensitivity of causal claims, invariantist or contextualist. Causal context sensitivity paddles, waddles, and quacks like semantic contrast sensitivity. But where are the ducks in the semantics?[16]

NOTES

1. Causal selection is often assimilated to the context sensitivity of 'the cause.' But these phenomena should be distinguished. On the one hand, multiple factors may be selected. The engineer, for instance, might select both the presence of the potholes and the absence of a stop sign as causes of the accident. On the other hand, the context sensitivity of 'the cause' is at least partly a matter of the separate context sensitivity of 'the' and does not obviously have anything to do with 'cause' (any more than the context sensitivity of 'the dog' automatically establishes any context sensitivity for 'dog').

2. Anscombe (1969) provides a similar example. She notes that one might accept that de Gaulle's making a speech caused an international crisis, but deny that the man with the biggest nose in France's making a speech caused an international crisis (without denying the facts). One wants to say: the size of the nose was not relevant. But unless one has an implausibly fine conception of events, there was only one speech.

3. While it is theoretically possible to reject both *Causal Contextualism* and *Causal Invariantism* (for instance by rejecting the "data" of the previous section), I am not aware of any theorists who have taken this approach. One theoretically alternative that does come up is to treat 'cause' as semantically *ambiguous*, as per Davidson's suggestion that 'cause' is ambiguous between the relation of causation and the sentential connective of causal explanation (1980, 162). But—though I lack the space for a proper discussion of Davidson's suggestion—I do not think it withstands much scrutiny. To my knowledge no serious linguistic evidence for any such ambiguity has been mooted, nor is there any reason to think that the sensitivities illustrated are due to "disambiguation." Indeed the ambiguity claim should entail that all of the causal claims considered have true readings, which should thereby be favored by charity. But the data is rather that seemingly "true" causal claims—such as the claim in 3 that the presence of oxygen caused the forest fire—still count as *unacceptable* in certain contexts. So unless some interpretive pressures are revealed which might overturn charity, ambiguity claims just get the data wrong. (In general, ambiguity claims multiply opportunities to find acceptable interpretations, and so they are good for explaining acceptabilities but not so good for explaining unacceptabilities.)

4. Lewis does tolerate some semantic context sensitivity in causal discourse, both with respect to the vagueness of counterfactuals ("The vagueness of similarity does infect causation, and no correct analysis can deny it" (1986, 163)) and—in his later influence account (2000)—with respect to the degree of influence sufficient for counting as a cause. But these look to be independent from the context sensitivities in section 1.1. So Lewis is perhaps best classified as a friend of *Causal Contextualism*, but one who sides with *Causal Invariantism* with respect to the issues under consideration in the main text.

5. I work with the Gricean view of conversational pragmatics simply because it is the most orthodox and developed approach. The invariantist who prefers a different view of pragmatics should take the discussion in the main text as a challenge to do better.

6. For discussions connecting the Gricean maxim of Relevance to the question under discussion ("speak to the question!") see Roberts 2004.

7. Here I am generalizing an argument due to McGrath, who considers pragmatic explanations for why we deny that certain omissions are causes (for instance, why we deny that the queen of England's failing to water my

flowers caused them to wilt), and points out: "it isn't just that we refuse to utter [omission sentences] that are, on the view, true; we also utter their negations" (2005, 128–129). Similar points arise in the literature on epistemic contextualism, as brought out by DeRose (1999).

8. Other advocates of the view that causal reasoning involves the notion of deviation from a default include Maudlin (2004) and McGrath (2005). But for Maudlin the notion of a default is encoded in the laws of nature (to the extent it is recoverable at all), and for McGrath the notion of a default comes from our notion of what is normal. As far as I can see, neither explicitly allows for context sensitivity, although both certainly could. Indeed McGrath's notion of what is normal strikes me as most naturally understood as a context-sensitive notion. What counts as "normal" for the forest rangers may be quite different from what counts as "normal" for the visiting Venusians.

9. This is perhaps clearest in Halpern's (2000) formalism for causal models, in which one begins from a *signature* <U, V, R>, where U is a set of *exogenous variables* ("initial conditions of the system"), V is a set of *endogenous variables* ("dependent conditions"), and R is a *range function* associated each variable $X \in U \cup V$ with a range of at least two allotted values. R encodes contrasts for the totality of events represented in the model. See also Eagle 2007 and Schaffer 2010.

10. Hall (2007) uses defaults to distinguish two sorts of causal structures which standard Pearl models conflate. To my mind this is the most promising case to be made for thinking that the notion of default is also essential to characterizing causal notions.

11. Arguably analogues of all of these sensitivities are to be found in knowledge ascriptions (Schaffer 2005b), with the seeming exception of selection effects. There does not seem to be anything on the epistemic side corresponding to selection (this would be a contextually variable tendency to promote certain elements from a subject's body of knowledge to the level of knowledge, and to demote the remaining elements to background information). This seeming disanalogy should be a mystery for everyone. For those who go in for parallel treatments of the context sensitivity of causal claims and knowledge ascriptions (such as myself), the mystery is why there is a minor *disanalogy*. For those who do not go in for parallel treatments, the mystery is why there is a *minor* disanalogy. Since my treatments of the context sensitivity of causal claims and knowledge ascriptions involve some minor differences (with knowledge claims there is only one contrast, with causal claims there are two), I should like to appeal to these minor differences to explain the minor disanalogy, but I know not how.

12. More precisely, Rooth adds a semantic focus marker whose value is a contextually determined set of options, and posits a dual interpretation of phrases with this marker. To illustrate, 'Socrates's DRINKING HEMLOCK at dusk' is semantically interpreted as [. . .[Socrates's [drinking hemlock]$_F$ at dusk]. . .], where [drinking hemlock]$_F$ induces a dual interpretation: there is the "ordinary semantic value" of drinking hemlock, and the "focus semantic value" which is the set of contextually salient options for what Socrates might have done at dusk (including drinking hemlock, but including other options as well). Semantic sensitivity to focus is then understood in terms of operators sensitive to these focus semantic values. Given that causal claims exhibit semantic sensitivity to focus, and given Rooth's alternative semantics for focus, it falls out that 'cause' is contrast sensitive.

13. *Lacuna*: if 3 does not receive any natural interpretation than its denial should not either, which does not quite fit that data in 13. So it would be smoother for me to say that 3 does receive some interpretation as a contrastive falsehood in

the context of the forest rangers. But I do not currently have any contrastive falsehood to suggest for the role.

14. Selection is the one aspect of context sensitivity that seems not to apply equally to both the cause and effect side, operating primarily on the cause side. There may also be something like selection on the effect side in our intuitive distinction between causes and byproducts, but this matter needs further exploration. (Selection seems to have a variety of special features, and may ultimately need separate treatment).

15. Cappelen and Lepore (2005) provide a useful battery of tests for indexicality (I think they mistake these for tests for context sensitivity generally, and don't properly consider the prospect that context sensitivity might come in multiple forms.)

16. Thanks to Mark Heller, Chris Hitchcock, Cei Maslen, and Peter Menzies.

REFERENCES

Achinstein, P. 1975. Causation, Transparency, and Emphasis. *Canadian Journal of Philosophy* 5: 1–23.

Anscombe, G.E.M. 1969. Causality and Extensionality. *Journal of Philosophy* 66: 152–159.

Bennett, J. 1995. *The Act Itself.* Oxford: Oxford University Press.

Cappelen, H., and E. Lepore. 2005. *Insensitive Semantics: A Defense of Semantic Minimalism and Speech Act Pluralism.* Oxford: Basil Blackwell.

Davidson, D. 1980. Causal Relations. In *Essays on Actions and Events,* 149–162. Oxford: Oxford University Press.

DeRose, K. 1999. Contextualism: An Explanation and Defense. In *The Blackwell Guide to Epistemology,* ed. J. Greco and E. Sosa, 187–205. Oxford: Basil Blackwell.

Eagle, A. 2007. Pragmatic Causation. In *Causation, Physics, and the Constitution of Reality: Russell's Republic Revisited,* ed. H. Price and R. Corry, 156–90. Oxford: Oxford University Press.

Garfinkel, A. 1981. *Forms of Explanation: Rethinking the Questions in Social Theory.* New Haven: Yale University Press.

Groenendijk, J., and M. Stokhof. 1984. Questions. In *Handbook of Logic and Language,* ed. J. van Bentham and A. ter Meulen, 1055–1124. Amsterdam: Elsevier.

Hall, N. 2007. Structural Equations and Causation. *Philosophical Studies* 132: 109–136.

Halpern, J. 2000. Axiomatizing Causal Reasoning. *Journal of Artificial Intelligence Research* 12: 317–337.

Halpern, J.Y., and J. Pearl. 2005. Causes and Explanations: A Structural-Model Approach. Part 1: Causes. *British Journal for the Philosophy of Science* 56: 843–887.

Hart, H.L.A., and A.M. Honoré. 1985. *Causation in the Law.* Oxford: Oxford University Press.

Hitchcock, C. 1996. The Role of Contrast in Causal and Explanatory Claims. *Synthese* 107: 95–419.

Hitchcock, C. 2003. Of Humean Bondage. *British Journal for the Philosophy of Science* 54: 1–25.

Hitchcock, C. 2007. Prevention, Preemption, and the Principle of Sufficient Reason. *Philosophical Review* 116: 495–532.

Kierkegaard, S. 1978. *Parables of Kierkegaard,* ed. T.C. Oden. Princeton: Princeton University Press.

Lewis, D. 1986. *Philosophical Papers*. Vol. 2. Oxford: Oxford University Press.

Lewis, D. 2000. Causation as Influence. *Journal of Philosophy* 97: 182–197.

Mackie, J.L. 1974. *The Cement of the Universe*. Oxford: Oxford University Press.

Maslen, C. 2004. Causes, Contrasts, and the Nontransitivity of Causation. In *Causation and Counterfactuals*, ed. J. Collins, N. Hall, and L.A. Paul, 341–357. Cambridge, MA: MIT Press.

Maudlin, T. 2004. Causation, Counterfactuals, and the Third Factor. In *Causation and Counterfactuals*, ed. J. Collins, N. Hall, and L.A. Paul, 419–443. Cambridge, MA: MIT Press.

McDermott, M. 1995. Redundant Causation. *British Journal for the Philosophy of Science* 40: 523–544.

McGrath, S. 2005. Causation by Omission: A Dilemma. *Philosophical Studies* 123: 125–148.

Menzies, P. 2004. Difference-Making in Context. In *Causation and Counterfactuals*, ed. J. Collins, N. Hall, and L.A. Paul, 139–180. Cambridge, MA: MIT Press.

Menzies, P. 2007. Causation in Context. In *Causation, Physics, and the Constitution of Reality*, ed. H. Price and R. Corry, 191–223. Oxford: Oxford University Press.

Mill, J.S. 1950. *A System of Logic*. New York: Macmillan.

Partee, B. 1989. Binding Implicit Variables in Quantified Contexts. *Proceedings of the Chicago Linguistics Society* 25: 342–365.

Pearl, J. 2000. *Causality: Models, Reasoning, and Inference*. Cambridge: Cambridge University Press.

Putnam, H. 1982. Why There Isn't a Ready-Made World. *Synthese* 51: 141–167.

Roberts, C. 2004. Context in Dynamic Interpretation. In *The Handbook of Pragmatics*, ed. L. Horn and G. Ward, 197–220. Oxford: Basil Blackwell.

Rooth, M. 1992. A Theory of Focus Interpretation. *Natural Language Semantics* 1: 75–116.

Schaffer, J. 2005a. Contrastive Causation. *Philosophical Review* 114: 327–358.

Schaffer, J. 2005b. Contrastive Knowledge. *Oxford Studies in Epistemology* 1: 235–271.

Schaffer, J. 2010. Contrastive Causation in the Law. *Legal Theory* 16: 259–297.

Schaffer, J. *forthcoming*. Necessitarian Propositions. *Synthese*.

Sperber, D. and D. Wilson. 1986. *Relevance: Communication and Cognition*. Oxford: Basil Blackwell.

Stanley, J. 2000. Context and Logical Form. *Linguistics and Philosophy* 23: 391–434.

Van Fraassen, B. 1980. *The Scientific Image*. Oxford: Oxford University Press.

Woodward, J. 2003. *Making Things Happen: A Theory of Causal Explanation*. Oxford: Oxford University Press.

3 Contrastive Bayesiansim

Branden Fitelson

1 WHAT *IS* "BAYESIANISM"?

I.J. Good (1971) once estimated that there are 46,656 varieties of Bayesianism. He based his estimate on a number of "dimensions" along which different sorts of Bayesianism could be characterized. From the perspective of this volume, Good's is probably an *under*estimation, because he was talking mainly about applications of "Bayesianism" to problems involving statistical inference. In contemporary analytic philosophy, there are still further "dimensions" along which (many) additional "Bayesianisms" might be distinguished. What all "Bayesianisms" have in common is that they all make essential use of *probability* as their main theoretical tool. For the most part, disagreements among different kinds of "Bayesians" will involve differing *interpretations* of probability. There are many interpretations of probability in the philosophical universe (see Hajek 2010 for an excellent survey). In this article, I will try to remain as neutral as possible on the various interpretive disputes that arise among the myriad "Bayesians" one encounters in the philosophical literature. Of course, certain applications will most naturally be associated with certain kinds of "Bayesianism" (in contrast with others). But, I will not dwell on such differences, unless they are essential to the "contrastivist" character of the accounts in question. Because my focus will be rather narrow (I'll be focusing on issues that arise in recent applications of Bayesian confirmation theory), I will be able to sidestep many (but, as we will see, not quite all) of these intramural interpretive disputes.

2 WHAT *IS* "CONTRASTIVISM"?

This volume is about "contrastivism." So, it is natural to wonder what distinguishes "contrastive" philosophical approaches or accounts from "noncontrastive" ones. I won't attempt a demarcation. Indeed, I won't even say very much (in any *general* way) about this distinction. Instead, I will try to illustrate how "contrastivist" thinking arises in some recent applications of "Bayesian" techniques. Hopefully, this will give some sense of how

"Bayesian philosophers" (broadly construed) think about "contrastivism" and its philosophical significance. To this end, I will examine several recent case studies from the contemporary literature on Bayesian confirmation theory, which, as we shall see, is implicated in Bayesian philosophy of science, Bayesian epistemology, and Bayesian cognitive science.

3 LIKELIHOODISM, BAYESIANISM, AND CONTRASTIVE CONFIRMATION

It is useful to begin with a discussion of a prominent "contrastivist" probabilistic account that has appeared in contemporary philosophy of science. This will simultaneously set the theoretical stage for subsequent sections, and illustrate a concrete example of "contrastivism" in (broadly) Bayesian philosophy of science.

Elliott Sober has been defending what he calls "contrastive empiricism" (CE) for over twenty years. In his original statement and defense of (CE), Sober (1994, p. 123) explains:

> Theory testing is a contrastive activity. If you want to test a theory *T*, you must specify a range of alternatives—you must say what you want to test *T against*. There is a trivial reading of this thesis that I do not intend. To find out if *T* is plausible is simply to find out if *T* is more plausible than *not-T*. I have something more in mind: there are various contrasting alternatives that might be considered. If *T* is to be tested against *T'*, one set of observations [*E*] may be needed, but if *T* is to be tested against *T''* a different set of observations [*E'*] may be needed. By varying the contrasting alternatives, we formulate genuinely different testing problems.

Here, Sober intends to be contrasting his view of the testing of scientific theories [(CE)] with what he takes to be its main philosophical rival: *Bayesian confirmation theory* (BCT). As Carnap (1962) explains, there are two distinct probabilistic notions of "support" or "confirmation" that "Bayesians" (broadly construed) must be careful to distinguish (I will be returning to this crucial distinction in subsequent sections):

- Confirmation as firmness (confirms$_f$). *E* confirms$_f$ *H* (relative to background corpus *K*) iff $\Pr(H \mid E \ \& \ K) > t$, where *t* is some (possibly contextually determined) threshold value. And, the *degree* to which *E* confirms$_f$ *H*, relative to background corpus *K*. [$c_f(H, E \mid K)$] is given by $\Pr(H \mid E \ \& \ K)$.
- Confirmation as increase in firmness (confirms$_i$). *E* confirms$_i$ *H* (relative to background corpus *K*) iff $\Pr(H \mid E \ \& \ K) > \Pr(H \mid K)$. And, the *degree* to which *E* confirms$_i$ *H*, relative to background corpus *K*

[$c_i(H, E \mid K)$], is given by some function [c_i] of $\Pr(H \mid E \& K)$ and $\Pr(H \mid K)$, where (informally) c_i is some measure of "the degree to which evidence E increases the firmness/probability of H (relative to background corpus K)."[1]

Intuitively, $c_f(H, E)$ is a measure of *how probable H is, on the supposition of evidence E*.[2] This is often called the *posterior* probability of H *on evidence E*. But, $c_i(H, E)$ is something different. It's meant to be a measure of the degree to which E is *evidentially relevant* to H, where this is explicated in terms of (some measure of) the degree to which E is *probabilistically relevant* to H. To gauge $c_i(H, E)$, we must *compare* the *posterior* probability of H *on E* with the *prior* probability of H.

On closer inspection, we can see that both kinds of confirmation-theoretic notions are implicated in Sober's quotation, above. Sober talks about a hypothesis H being "plausible," which he takes to be synonymous with H's being "more plausible than *not-H*." Here, Sober has in mind the *firmness* concept c_f. He is talking about a hypothesis H being "more probable than not" (given E). Formally, he has in mind cases in which $\Pr(H \mid E) > \Pr(\sim H \mid E)$, which is equivalent to $\Pr(H \mid E) > 1/2$, *i.e.*, $c_f(H, E) > 1/2$. Sober wants to distinguish this "non-contrastive" plausibility claim with a "contrastive" claim to the effect that evidence E *favors* one hypothesis H_1 over a *concrete alternative* hypothesis H_2 (where H_2 is *not* equivalent to *not-H_1*).

When it comes to the *contrastive* "favoring" relation, Sober (1994) is a *Likelihoodist*. That is, he accepts the following *Law of Likelihood* (for all relations of favoring).

(LL) Evidence E favors hypothesis H_1 over hypothesis H_2 *if and only if* H_1 confers greater probability on E than H_2 does.

[Formally, E favors H_1 over H_2 iff $\Pr(E \mid H_1) > \Pr(E \mid H_2)$.]

This is called the "Law of Likelihood" because it says that the relation "E favors H_1 over H_2" boils down to a comparison of the *likelihoods* $\Pr(E \mid H_1)$ and $\Pr(E \mid H_2)$ of the two alternative hypotheses H_1 and H_2, respectively— *relative to evidence E*.

Interestingly, Sober's Likelihoodist *favoring* relation is intimately related to Carnap's *confirmation as increase in firmness* concept c_i. To see this, we need to delve a bit deeper into c_i-theory. First, note that various c_i-measures have been proposed and defended by (BCT)-ers. Here are a few of the most popular c_i-measures.[3]

- *Difference*: $d(H, E) =_{df} \Pr(H \mid E) - \Pr(H)$
- *Ratio*: $r(H, E) =_{df} \log\left[\frac{\Pr(H \mid E)}{\Pr(H)}\right]$
- *Likelihood-Ratio*: $l(H, E) =_{df} \log\left[\frac{\Pr(E \mid H)}{\Pr(E \mid \sim H)}\right] = \log\left[\frac{\Pr(H \mid E) \cdot [1 - \Pr(H)]}{[1 - \Pr(H \mid E)] \cdot \Pr(H)}\right]$

Second, consider the following *bridge principle*, which connects the *favoring* relation and certain comparative claims involving *confirmation as increase in firmness*:

($\dagger_{c_i}$) Evidence E favors hypothesis H_1 over hypothesis H_2, according to measure c_i, *if and only if* $c_i(H_1, E) > c_i(H_2, E)$.

What ($\dagger_{c_i}$) says is that *favoring* relations *supervene on c_i-relations*. This is a natural bridge principle for a (BCT)-theorist to accept. After all, it just says that E favors H_1 over H_2 iff E confirms$_i$ H_1 more strongly than E confirms$_i$ H_2. I will discuss the philosophical case for (and against) such a bridge principle, below. But, first, note that ($\dagger_{c_i}$) forges the following intimate connection between (LL) and (BCT):

(1) ($\dagger_r$) entails (LL).[4]

What (1) says is that if one adopts *both* (i) r as one's c_i-measure, *and* (ii) the bridge principle ($\dagger_{c_i}$) connecting favoring and comparative-c_i, then one's (Bayesian) confirmation$_i$-theory *entails* the Law of Likelihood (LL). Moreover, it can be shown that r is the *only* choice of c_i-measure (among contemporary c_i's) that entails [*via* ($\dagger_{c_i}$)] the (LL). In this sense, it is somewhat misleading for Sober to represent (BCT) and (LL) as *mutually exclusive* alternative approaches to "favoring." In fact, what (1) reveals is that Sober's (LL) is a *consequence of a particular way of being a Bayesian confirmation-theorist*. And, as it happens, one prominent Bayesian (Milne 1996) has *used* ($\dagger_{c_i}$) and (LL) to argue in favor of r as "the one true measure of c_i."

Thus, we find ourselves in a somewhat uncomfortable dialectical position. We have both Likelihoodists and (some) Bayesians accepting the Law of Likelihood (LL), but (all) Likelihoodists seem to think that their approach is (somehow) *incompatible* with *any* Bayesianism. I suspect that the problem here has to do with the *bridge principle* ($\dagger_{c_i}$). Whereas it would be natural for a *Bayesian* (such as Milne) to accept ($\dagger_{c_i}$), it is not at all clear why a *Likelihoodist* (*per se*) should accept ($\dagger_{c_i}$). Indeed, it seems to me that Likelihoodists must *reject* ($\dagger_{c_i}$), *if* they are to claim that their theory of favoring is somehow *incompatible* with a (BCT) that accepts ($\dagger_r$).

But, this is a delicate matter. After all, a proponent of the ($\dagger_r$)-flavor of (BCT) will *agree* with Likelihoodists on *all favoring claims*. That is, there won't be any concrete examples involving hypothesis testing on which the Likelihoodist and the ($\dagger_r$)-theorist will disagree about any of the favoring relations. So, what *can* they disagree about? It seems that all they *can* disagree about is the *existence* of the "non-contrastive" confirmational quantities—$r(H_1, E)$ and $r(H_2, E)$—that feature in the Bayesian bridge principle ($\dagger_r$). But, why would Likelihoodists doubt the *existence* of $r(H_1, E)$ and $r(H_2, E)$? Likelihoodists often complain that the existence of such quantities presupposes the existence of *prior probabilities* $\Pr(H_1)$ and $\Pr(H_2)$

of the hypotheses H_1 and H_2. And, Likelihoodists are skeptical about the existence of prior probabilities of hypotheses. This has lead some Likelihoodists to view (LL) as a *"ceteris paribus* law." For instance, Sober (2006, 10) now says:

> The Law of Likelihood should be restricted to cases in which the probabilities of hypotheses are not under consideration (perhaps because they are not known or are not even "well defined") and one is limited to information about the probability of the observations given different hypotheses.

The worry here seems to trade on the fact that $r(H,E)$ is a function $\left[\frac{Pr(H\,|\,E)}{Pr(H)}\right]$ that features the prior probability of H, $\Pr(H)$, as a term (and Likelihoodists are skeptical about the existence of such priors). But, this is potentially misleading, because $r(H,E)$ is *numerically identical* to a function $\left[\frac{Pr(E\,|\,H)}{Pr(E)}\right]$, which does *not* feature the prior probability of H as a term—it features only the *likelihood* of H [$\Pr(E \mid H)$] and the prior probability of the *evidence E* [$\Pr(E)$]. So, being more careful now, I suppose Likelihoodists (for example, Sober) should say that they are *only* willing to countenance *likelihoods* of hypotheses $\Pr(E \mid H)$, and that they are skeptical about the existence of *prior probabilities of both hypotheses and bodies of evidence*.

What do we *mean* when we say that Likelihoodists are "skeptical about the existence of some conditional (see note 1) probabilities (for example, priors), but not others (for example, likelihoods)"? This is where we enter into vexed controversies involving various *interpretations* of the salient confirmation-theoretic conditional probability functions $\Pr(\bullet \mid \bullet)$. A Bayesian will typically interpret such conditional probabilities either as the *actual* (conditional) *degrees of belief* of an agent (Finetti 1989; Ramsey 1931) *or* as *justified* or *epistemically rational* (conditional) degrees of belief, relative to some body of evidence (Keynes 1921; Carnap 1962; Williamson 2002, ch. 10). Likelihoodists, on the other hand, are (mainly) talking about *statistical* probabilities, which are implied by *statistical hypotheses* (Royall 1997). And, when it comes to statistical probabilities, it is typically assumed that these are (paradigmatically) *likelihoods of statistical hypotheses*. Prior probabilities (of *either* hypotheses *or* bodies of evidence) are usually *not* determined by the sorts of statistical models Likelihoodists have in mind. As such, Likelihoodists seem to be assuming that *only* probabilities that are implied by certain sorts of statistical models are "fair game" for use in theories of favoring (or confirmation). In other words, Likelihoodists might be willing to grant that there is a *sense* in which (for example) subjective Bayesian probabilities "exist" (that is, *in the minds of Bayesians*), but Likelihoodists will maintain that such probabilities lack *epistemological probative value* in hypothesis testing (specifically, in the testing of *statistical* hypotheses).[5]

This is the best explanation of the motivation behind Sober's proposed *restriction* of (LL) that I have been able to come up with. But, I remain unconvinced. First, the restriction seems rather *ad hoc*. And, second, it seems to presuppose that—in whatever contexts the Likelihoodists have in mind—there aren't any *objective* (and probative) *epistemological constraints* on "initial credences" in hypotheses (and/or bodies of evidence). I'm not sure why we should believe that.[6] Having said that, I want to avoid getting bogged down in disputes involving competing interpretations of confirmation-theoretic conditional probabilities. For this reason, I won't delve any further into this particular dialectical labyrinth (but, see note 6 for references). Instead, I want to return to the *material adequacy* of the so-called "Law of Likelihood." I think the (LL)—in its original formulation—is simply *false*.

Here is what I take to be a rather clear counterexample to (LL). The experimental set up for my counterexample to (LL) involves a standard, well-shuffled deck of playing cards, from which we are going to sample a single card. Let $E =_{df}$ the card is a spade, $H_1 =_{df}$ the card is the ace of spades, and $H_2 =_{df}$ the card is black. In this example (assuming the standard statistical model of card draws), we have $\Pr(E \mid H_1) = 1 > \Pr(E \mid H_2) = 1/2$. So, according to (LL), E favors H_1 over H_2. But, this seems absurd. After all, the truth of E *guarantees* the truth of H_2, but the truth of E does *not* guarantee the truth of H_1. In this sense, E constitutes *conclusive evidence* for H_2, and *less than* conclusive evidence for H_1. If *that* doesn't imply that E *favors* H_2 over H_1, then what would? This suggests the following principle:

(CE) If E constitutes conclusive evidence for H_1, but E constitutes less than conclusive evidence for H_2 (where it is assumed that E, H_1, and H_2 are all contingent), then E favors H_1 over H_2.

It seems to me that (CE) should be a *desideratum* for any adequate explication of "favoring." And, because (LL) implies the existence of counterexamples to (CE), this seems to *refute* (LL). Moreover, (CE) has strong ramifications for any Bayesian theory of confirmation, which accepts the bridge principle ($\dagger_{c_i}$). Once we accept (CE) and ($\dagger_{c_i}$), then we must *not* adopt r as our measure of c_i, because ($\dagger_r$) entails (LL).

One way to respond to this counterexample would be to adopt Sober's restriction of (LL) to contexts in which "only likelihoods are available, and no priors are available." However, it is unclear whether the intuitive verdict about this example even *depends on* the "availability of priors." It seems to me that there is a simple *logical asymmetry* that explains the intuitive verdict that E favors H_2 over H_1 in the example. I don't think this requires any appeal to "verboten priors." As a result, I don't think the Soberian "*ceteris paribus* law" reading of (LL) is helpful here.

Another way to respond to my counterexample to (LL) would be to find a different strategy for "restricting the scope" of (LL)—one which is motivated in some other way. Recently, Dan Steel (2007) and Jake Chandler (2010) have (independently) responded to (CE) and my putative counterexample to (LL) in just such a way. Both Steel and Chandler argue that "favoring" is inherently *contrastive* in nature. And, as a result, they argue that the hypotheses H_1 and H_2 that appear in (LL) and (CE) must be *mutually exclusive*. This rules out my example, because in my example H_1 entails H_2.[7]

Following Hitchcock's approach to contrastive probabilistic explanation (1996, 1999)—which presupposes that members of contrast classes (of *explananda*) are *mutually exclusive*—Chandler (2007, 2010) proposes the following Hitchcock-style account of favoring that *builds in* mutual exclusivity of the alternative hypotheses:

> (HC) E favors H_1 over H_2 iff (i) H_1 and H_2 are mutually exclusive, and (ii) $\Pr(H_1 \mid E \ \& \ (H_1 \vee H_2)) > \Pr(H_1 \mid {\sim}E \ \& \ (H_1 \vee H_2))$.

Interestingly, it turns out that (HC) is *logically equivalent* to the following:

> (LL*) E favors (LL*) H_1 over H_2 iff (i) H_1 and H_2 are mutually exclusive, and (ii) $\Pr(E \mid H_1) > \Pr(E \mid H_2)$.

That is, assuming mutual exclusivity of H_1 and H_2, the Hitchcock-Chandler approach to favoring (HC) *is equivalent to the Likelihoodist theory of favoring* (LL).[8]

This suggests a natural revision/restriction of the original Law of Likelihood (LL), which is embodied in (LL*), above. The idea is that (LL) is true—*provided that the alternative hypotheses H_1 and H_2 are mutually exclusive*. This restriction of (LL) is *not ad hoc* in the way that Sober's restriction was. Indeed, the restriction does have some intuitive plausibility. When we make contrastive claims, we *often* presuppose that the members of the salient contrast class are mutually exclusive. But, it is natural to ask whether such a presupposition is *always* present.

It seems clear to me that such a presupposition is *not always* present in cases of favoring/contrastive confirmation.[9] For instance, in the context of statistical hypothesis testing, Likelihoodists (and other "anti-Bayesians") are quick to *criticize* Bayesian approaches that *do* (generally) presuppose the mutual exclusivity of alternative hypotheses. Indeed, in the context of statistical model selection, it is supposed to be one of the relative *strengths* of Likelihoodism—*as opposed to* certain flavors of Bayesianism—that it *is* capable of testing *nested models* (that is, models that bear *containment or entailment* relations to each other, as my H_1 and H_2, above, do) against each other. So, this "mutual exclusivity requirement" reply to my counterexample to (LL) is simply *not available* to traditional, *statistical* Likelihoodists. Structurally analogous cases involving statistical model selection will be among the examples *used by* Likelihoodists as "evidence" that is supposed to "favor" their own approach to model selection over certain

Bayesian approaches. See (Forster 1995) for a thorough discussion of this debate among philosophers of statistics.

Another worry I have about the "mutual exclusivity presupposition" maneuver is that it seems to make it *too easy* to refute the bridge principle ($\dagger_{c_i}$). A Bayesian confirmation theorist who accepts a bridge principle ($\dagger_{c_i}$) will hear "E favors H_1 over H_2" as *synonymous* with "E evidentially supports H_1 more strongly than E evidentially supports H_2." And, it seems clear that this latter *comparative confirmation* claim does *not* (always) presuppose that the alternative hypotheses (H_1 and H_2) are mutually exclusive. Surely, evidence E can sometimes support a logically stronger (or logically weaker) hypothesis more strongly than E supports a logically weaker (or logically stronger) one. In the next two sections, I will discuss two well-known illustrations of this phenomenon. This will provide further test cases for the various accounts of "favoring" (and confirmation) that we've been discussing.

4 THE PROBLEM OF IRRELEVANT CONJUNCTION

This section has three parts. In the first part, I will rehearse (in some detail) the historical dialectic concerning the "problem of irrelevant conjunction" (PIC). In the second part, I will explain how this problem (and our proposed resolution of it) can shed some light on the questions about Likelihoodism, Bayesianism, "favoring," and "contrastive confirmation" that were discussed in the previous section. In the third part of this section, I will discuss an objection to our approach to (PIC) that was recently raised by Patrick Maher, and an alternative, "contrastivist" account of (PIC) due to Jake Chandler, which was inspired by Maher's objection.

4.1 Part I: The Historical Dialectic of the Problem of Irrelevant Conjunction

One of the traditional (deductive) accounts of confirmation that features prominently in the history of confirmation theory is the so-called Hypothetico-Deductive (or HD) account of confirmation (I will call this confirmation relation "confirms$_h$"):

(HD) E confirms$_h$ H iff H entails E.

Due to the monotonicity of entailment, confirmation$_h$ has the following property:

(2) If E confirms H, then E confirms H & X (for *any* X).

Clark Glymour (1980, p. 31) raises two worries in connection with property (2):

(2a) If H entails E, then so will H & X, where X is any sentence whatsoever. But, we cannot admit, generally, that E will lend any plausibility
to an arbitrary X. One might, of course, deny what Hempel calls the
special consequence condition, namely, that if E confirms a hypothesis, then E will confirm every logical consequence of that hypothesis.
But this is hardly satisfactory. Sometimes, anyway, confirmation does
follow entailment, at least over some paths.

(2b) As evidence accumulates, we may come to accept a hypothesis . . .
and when we accept a hypothesis we commit ourselves to accepting
all of its logical consequences. So, if a body of evidence could bring
us to accept hypothesis H, and whatever confirms H confirms H &
X, where X is any irrelevant hypothesis, then the same evidence that
brings us to accept H, ought, presumably, to bring us to accept X.

Both of these worries about (2) have to do with (some of) the evidential
support provided by E (for H) somehow "rubbing off" onto an *irrelevant
conjunct* X. But, the two worries seem to involve *two different notions* of
evidential support:

- E **supports**$_1$ H iff E is (positively) evidentially relevant to H.
- E **supports**$_2$ H iff E warrants/justifies belief/acceptance of H.

Glymour's worry (2b) involves support$_2$, and Glymour's worry (2a)
involves support$_1$. As I mentioned above, Carnap cautioned us not to
conflate confirms$_f$ and confirms$_i$. Similarly, we need to be careful not to
conflate these two notions of "evidential support." For the sake of the
present discussion, we will follow Carnap, who thought of confirms$_i$ as
an *explicatum* for supports$_1$, and confirms$_f$ as an *explicatum* for sup-
ports$_2$. Thus, we will say that Glymour's worry (2a) involves confirms$_i$,
and his worry (2b) involves confirms$_f$. Finally, *nobody* thinks that con-
firms$_h$ is a good *explicatum* for supports$_2$.[10] So, I won't bother to discuss
Glymour's worry (2b).

Glymour's (2a) is more interesting. This worry has become known as
"the problem of irrelevant conjunction" (PIC). In worry (2a), Glymour
mentions Hempel's *special consequence condition* (Hempel 1945) which
entails the following condition:

(SCC) If E confirms H & X, then E confirms X (for *any* X).

Clearly, no adequate account of confirmation can satisfy *both* (2) *and*
(SCC). Any such theory would entail that *any* evidence E (if it confirms
any hypothesis) will confirm *every* proposition X. This is why Glymour
suggests that one "cheap" way out of the problem of irrelevant conjunction
is to (merely) *deny* (SCC). But, as Glymour also suggests, we want *more
than a mere denial of* (SCC) here. We want a response to "the problem of

irrelevant conjunction" that also involves a *principled* (and intuitively plausible) way of determining when (SCC) holds and when it fails.

This is where Bayesian confirmation theory (namely, confirmation$_i$-theory) comes into the picture. First, note that confirmation$_i$-theory does *not* entail (2). That is:

(3) E confirms$_i$ H does *not* entail E confirms$_i$ H & X.

But, confirmation$_i$-theory *does* entail the following *deductive special case* of (2):

(4) If H *entails* E, then E confirms$_i$ H & X (for *any* X).

Fact (4) has inspired several confirmation$_i$-theorists to offer "resolutions" of (PIC). John Earman (1992) appeals to the following to try to "soften the impact" of (4):

(4.1) If H entails E, then $\mathbf{c}_i$ $(H$ & $X, E) < \mathbf{c}_i$ (H, E).

The idea behind Earman's (4.1) is that—although it is true that confirms$_i$ entails the (PIC)-like property (4) in the (HD) case where H entails E—tacking irrelevant conjuncts onto such [(HD)-confirmed] hypotheses will always *lower the degree to which E confirms$_i$ them.* That is, one will *pay a confirmation$_i$-theoretic price* for tacking irrelevant conjuncts onto an [(HD)-confirmed] hypotheses. Closer scrutiny of Earman's response to (4) reveals the following three features:

(a) The "irrelevance" of the conjuncts X is *irrelevant* to the decrease in $\mathbf{c}_i$. After all, (4.1) is true for *all* X—irrelevant or otherwise. It would be preferable if the irrelevance of X was (in some sense) playing an explanatory role.
(b) (4.1) is *not true for all* measures $\mathbf{c}_i$ of confirmation$_i$. For instance, (4.1) fails to hold for the ratio measure $r(H, E)$ that we discussed above, in connection with Likelihoodism. As such, proponents of (†$_r$), for example, Milne, won't be in a position to avail themselves of Earman's approach to (PIC).
(c) (4.1) only applies to cases of *deductive* evidence (that is, cases in which E confirms$_h$ H). As we'll see shortly, confirmation$_i$ faces a *more general* (PIC)-type problem. And, Earman's approach won't be applicable to it.

Other Bayesians have offered similar responses to (PIC)/(4). Rosenkrantz (1994) appeals to the following, rather similar result:

(4.2) If H entails E, then $d(H$ & $X, E) = \Pr(X \mid H) \cdot d(H, E)$.

Using (4.2), Rosenkrantz tries to address *some* of the problems we raised for Earman's (4.1)-approach. Rosenkrantz explains the rationale behind (4.2) as follows:

> I hope you will agree that the two extreme positions on this issue are equally unpalatable, (i) that a consequence E of H confirms H & X not at all, and (ii) that E confirms H & X just as strongly as it confirms H alone. . . . In general, intuition expects intermediate degrees of confirmation that depend on the degree of compatibility of H with X. (Rosenkrantz 1994, 471)

Basically, what Rosenkrantz is doing here is: (i) adopting $Pr(X \mid H)$ as a measure of the "degree of compatibility of H with X," and (ii) adopting the difference measure d as his measure of confirmation$_i$. Does this ultimately lead to an *improvement* on Earman's (4.1)? This depends on whether Rosenkrantz really has adequately addressed worries (a)–(c), above. Unfortunately, I don't think he has.

In a way, Rosenkrantz is *trying* to address (a) here. He seems to be thinking of $Pr(X \mid H)$ as a kind of measure of "the degree of (ir)relevance" of X—*qua conjunct* in H & X. But, this is a *peculiar* way for a *Bayesian* to explicate "(ir)relevance." Normally, Bayesians use *probabilistic (in)dependence* relations to explicate *(ir)relevance* relations. Moreover, because it is a relation involving only X and H, $Pr(X \mid H)$ can tell us nothing about "degrees of relevance" involving X, H—*and E.* And, in general, this "irrelevant conjunct" relation (whatever it turns out to be) must be *evidence relative*. That is, what we want is an explication of "X is an irrelevant conjunct *to H, relative to evidence E.*" When it comes to (b), Rosenkrantz is in even worse shape than Earman. Rosenkrantz's approach works *only* for confirmation$_i$-measures that are *very similar* to the difference measure d. In this sense, Earman's approach is strictly more general. Finally, Rosenkrantz is still only addressing the *deductive* case. So, like Earman's approach, Rosenkrantz's approach will not be useful for more general, inductive varieties of (PIC), which we will see shortly.

Ultimately, confirmation$_i$-theorists need to *rethink* the problem of irrelevant conjunction, and its possible resolution(s). To that end, let's think about how c_i-theory handles *irrelevant* conjunctions, in the general, *inductive* case. First, we need to say what it *means* for X to be an *irrelevant conjunct* to a hypothesis H, *with respect to evidence E.* Here, we can adopt stronger or weaker explications of "irrelevant conjunct." The strongest, natural Bayesian explication is:

- X is a *strongly irrelevant conjunct to H, with respect to evidence E,* just in case X is *probabilistically independent* of H, E, and H & E.[11]

One could also adopt the following weaker explication of "irrelevant conjunct":

- X is a *weakly irrelevant conjunct to H, with respect to evidence E*, just in case $\Pr(E \mid H \,\&\, X) = \Pr(E \mid H)$ [that is, *if H screens-off E from X*].

The idea behind the weak explication of "irrelevant conjunct" is that a weakly irrelevant conjunct X *does not affect the likelihood of the hypothesis H* (on evidence E). In this sense, X does not add anything to H—insofar as its predictions about (the probability of) the evidence are concerned. This is a natural (weak, Bayesian) way of capturing the idea that X is an "irrelevant conjunct" to H, with respect to evidence E. The strong explication entails the weak explication, but it also entails *much more*. Whereas the weak explication appeals *only to likelihoods* of hypotheses, the strong explication trades in *priors of both hypotheses and evidence*, and so will, presumably, not be something that Likelihoodists will (generally) find kosher.[12]

Now that we have both strong and weak explications of "irrelevant conjunct," we are in a position to investigate how tacking on irrelevant conjuncts affects confirmation relations (both qualitative and quantitative). Here are two key results:

(5) If E confirms$_i$ H, and X is an (either strongly or weakly) irrelevant conjunct to H, with respect to evidence E, then E *also* confirms$_i$ $H \,\&\, X$.

(6) If E confirms$_i$ H, and X is an (either strongly or weakly) irrelevant conjunct to H, with respect to evidence E, then $\mathbf{c}_i (H \,\&\, X, E) < \mathbf{c}_i (H, E)$, for all measures of $\mathbf{c}_i$ (under consideration), *except the ratio measure r*.

What (5) tells us is that confirmation$_i$-theory *does suffer from* a *general* problem of *irrelevant* conjunction, which is analogous to the problem faced by confirmation$_h$-theory (that is, HD-confirmation). That is, tacking *irrelevant* conjuncts onto a hypothesis that is confirmed$_i$ by E yields *conjunctions that are also confirmed$_i$ by E*. Result (6) is a generalization of Earman's (4.1), which avoids our criticisms (a) and (c) of Earman's account. Regarding (a), our (6) *makes essential use of* the *irrelevance* of the conjunct X. Regarding (c), we have generalized both the problem and its (Earman-style) resolution far beyond the deductive case in which H entails E. Indeed, we now see that the deductive case is just an *extreme, limiting-case* in which H *screens-off* X from E. The fact that *weak* irrelevance (namely, the screening-off of X from E, by H) is sufficient for both (5) and (6) provides a *unified explanation* of *why* the deductive *and* inductive cases of (PIC) behave the way they do. Regarding our criticism (b) of Earman's account, we can unfortunately do no better than Earman did. Alas, our results only go through for $\mathbf{c}_i$ measures *other than* r. This means that Bayesians (like Milne) who accept r will not be able to avail themselves of our approach to (PIC). And, those (like Milne) who accept the bridge principle ($\dagger_r$) will not be able to say that E *favors* H over $H \,\&\, X$ when X is an irrelevant conjunct to H, with respect to evidence E. This connection to ($\dagger_r$)—and therefore

(LL)—is an interesting and important one. I will return to this dialectical thread in Part II of this section. Before moving on to Part II of this section, however, it is useful to illustrate our account of (PIC) with the following simple concrete example:

- Suppose we'll be sampling a card at random from a standard deck. Let E be the proposition that the card is black. Let X be the hypothesis that the card is an ace, and let H be the hypothesis that the card is a spade.

In this example (assuming, as above, the standard probability model for random draws from a standard deck of playing cards), we have the following facts:

(7) E confirms, H.
(8) $\Pr(E \mid H \,\&\, X) = \Pr(E \mid H)$.

Thus, the preconditions of our (5) and (6) are met in the example. Therefore:

(9) E confirms, $H \,\&\, X$.
(10) c_i (H & X, E) < c_i (H, E), for all c_i, *except the ratio measure r*.

Finally, we *also* have the following important fact:

(11) E does *not* confirm, X.

In light of (9) and (11), this also constitutes a counterexample to (SCC) for confirms,. But, Bayesian confirmation, theory doesn't *merely reject* (SCC) here. Rather, it provides a principled and illuminating account of when (SCC) fails (and when it holds). This seems to satisfy Glymour's desire for an account of confirmation that (i) has something interesting and illuminating to say about (PIC), and (ii) simultaneously provides a principled and explanatory rejection of (SCC).[13] In Part II of this section, I return to Likelihoodism, favoring, and contrastive confirmation, in light of (PIC).

4.2   Part II: (PIC), Favoring, and Contrastive Confirmation

At the end of section 3, I mentioned that there were clear-cut examples in which evidence E supports H more strongly than E supports $H \,\&\, X$. In Part I of this section, I explained how (PIC) provides an interesting class of examples of precisely this kind, and how Bayesian confirmation, theory provides the (intuitively) correct verdicts concerning such cases. I also mentioned that advocates of the c_i measure r are unable to reproduce these verdicts. Here is a result that furnishes a more precise explanation:

(12) If E confirms, H, and X is an (either strongly or weakly) irrelevant conjunct to H, with respect to evidence E, then $r\ (H\ \&\ X,\ E) = r\ (H,\ E)$.

This is not surprising, because (i) "weak irrelevance" entails that the *likelihoods* of H and $H\ \&\ X$ are equal (relative to E), and (ii) a comparison of r $(H,\ E)$ and $r\ (H\ \&\ X,\ E)$ boils down to a comparison of the *likelihoods* of H and $H\ \&\ X$ (relative to E). Moreover, because $(\dagger_r)$ entails (LL), advocates of $(\dagger_r)$ will have to say that there can be *no favoring* of H over $H\ \&\ X$ by E, whenever X is an "irrelevant conjunct" (in either our weak or strong senses). What should a Likelihoodist say about this situation?

There are two ways that advocates of (LL) typically respond to these facts about (PIC). The first way [exemplified by Milne, and other "Bayesian-Likelihoodists" who adopt the bridge principle $(\dagger_r)$] is to *bite the bullet*, and insist that E *evidentially supports H and $H\ \&\ X$ equally strongly* in examples of "irrelevant conjunction" (like our concrete example involving the deck of cards in section 4.1, above). I won't address this strategy here (except to say that I don't find it intuitively compelling). Rather, I will focus on the second kind of response given by advocates of (LL).

The second type of response by defenders of (LL) comes from "contrastivist-Likelihoodists" (such as Chandler) who accept (LL), but do *not* accept the bridge principle $(\dagger_r)$. Such advocates of (LL) will insist that *it doesn't make sense* to talk about "favoring" in cases like (PIC), where the alternative hypotheses are *not* mutually exclusive. In other words, the second response is an instance of the *mutual exclusivity requirement* (on alternative hypotheses involved in favoring relations), which we discussed in section 3 above. Ultimately, I don't think this response obviates the need for something very much like Milne's bullet-biting response. This is because, as Susanna Rinard (2005) points out, requiring mutual exclusivity of alternative hypotheses doesn't manage to avoid all of the problems raised by (PIC)-type considerations. We can see Rinard's point clearly with this simple modification of our card-sampling example from section 4.1 (which is similar to her example):

- Let E be the proposition that the card is black. Let X be the hypothesis that the card is an ace, let H_1 be the hypothesis that the card is a spade, and let H_2 be the hypothesis that the card is a club.

"Contrastivist-Likelihoodists" (for example, Chandler) will complain that claims such as

(12) E *favors H_1 over $H_1\ \&\ X$*

[which, owing to (5) and (6), are implied by all bridge principles $(\dagger_{c_i})$ *except for* $(\dagger_r)$] are *infelicitous* on the grounds that "favoring" claims presuppose that the alternative hypotheses involved are mutually exclusive. However, mutual exclusivity is not really the issue here. To see this, consider the following *analogous* claim:

(13) E *favors* H_2 over H_1 & X.

Here, the alternative hypotheses *are* mutually exclusive. So, it seems that there can be no "constrastivist-presuppositional" grounds for claiming that (13) is infelicitous. Therefore, it seems that *all* advocates of (LL) must say that (13) is (felicitous and) *false*. After all, the *likelihoods* of H_2 and H_1 & X are *equal* (they are both equal to 1). Moreover, it seems clear that— from a probabilistic point of view—claims H_1 and H_2 *are on a par with respect to the evidential relations they bear to E and X* in our example. This seems to imply that something very much like Milne's *bullet-biting* response still needs to be embraced by defenders of (LL)—even those who reject the bridge principle $(\dagger_r)$ on "contrastivist-presuppositional" grounds. Because I think it's implausible to claim *either* that E evidentially supports H_2 and H_1 & X *equally strongly* or that E does *not* favor H_2 over H_1 & X in the present example, I don't hold out much hope for Likelihoodists (of either stripe) to tell a compelling, general story about (PIC). That said, I will briefly discuss a recent "contrastivist-Likelihoodist" alternative approach to (PIC), due to Chandler.

4.3   Maher's Objection and Chandler's Alternative "Contrastivist" (PIC)-Account

Patrick Maher (2004) has recently criticized our [(5)&(6)-based] approach to (PIC) (see Hawthorne and Fitelson 2004). He complains that our approach doesn't even *address* (PIC), because he thinks the (PIC) is *motivated essentially by the following (faulty) intuition about support*$_1$:

(*)   If X is an irrelevant conjunct to H, with respect to E, then E *does not* support$_1$ H & X.

Maher's approach to (PIC) is to simply explain why this is a *false intuition*. He does so by discussing concrete counterexamples to (*), much like our simple card examples in sections 4.1 and 4.2 above. Our response to Maher is quite simple. Of course, we agree that (*) is false (indeed—*that's part* of *our* story about (PIC), too!). But, we *disagree* with the claim that (PIC) is motivated (*essentially*) by "intuition (*)." We would say that (PIC) arises because of the *truth* of (5), above. It is (5) that implies the *existence* of a "problem of irrelevant conjunction" for supports$_1$/confirms$_i$. And, pointing out the *falsity* of (*)—which, of course, we *also* do—doesn't do anything to address (or "soften the impact") of the truth of (5). This is why we think (6) is an essential part of any complete Bayesian confirmation$_i$-theoretic story about (PIC). Thus, we think Maher has the wrong diagnosis here.[14]

Chandler (2007) has picked up on Maher's (*)-line on (PIC). He has taken Maher's line as a point of departure for his own "contrastivist-Likelihoodist" alternative to our (non-Likelihoodist) approach to (PIC). Chandler starts with his "contrastivist-Likelihoodist" approach to favoring: (HC)/(LL*), which (as we explained above) is just the Law of Likelihood, *plus* the requirement that alternative hypotheses in favoring relations must always be mutually exclusive. Then, he combines (HC)/(LL*) with Maher's claim that intuition (*) is what's driving (PIC). This culminates in the suggestion that people (falsely) *think* that the *non-contrastive* supports$_1$/confirms$_i$ claim (*) is true, because they are *conflating* it with a *true contrastive-favoring* claim. Specifically, Chandler suggests that such people are conflating (*) with

(**) If X is an irrelevant conjunct to H, with respect to evidence E, then E does not *favor* H & X over H & $\sim X$,

which, according to Chandler's contrastivist-favoring theory (HC)/(LL*), *is* true.

I'll just make a couple of brief remarks about Chandler's approach. First, as I've already explained, I think Maher's diagnosis of "intuition-(*)" as *the* (or even *a primary*) source of (PIC) is off the mark. But, let's bracket that and just focus on Chandler's discussion concerning (**). Here, I think his discussion is misleading in several ways. First, he presents his theory of favoring in its (HC)-form, which obscures the fact that it is equivalent to (LL*).[15] Second, once we realize that Chandler's theory is a *Likelihoodist* theory of favoring, it is a *trivial* matter that (**) comes out true on his theory of favoring, because he is assuming our weak explication of "irrelevant conjunction," which just *is* the salient Likelihood *identity*. Third, because (LL) is rather controversial, Chandler's claim that (**) is true is *also* rather controversial. Basically, *only* those who accept (LL) will accept (**). As a result, it is somewhat misleading of Chandler to represent his account as providing a "charitable reconstruction" of how someone might come to accept "something like" (*). Finally, it is often *psychologically implausible* to suggest that people attend to *contrastive* claims in contexts where they are confronted with *non-contrastive* questions (which is what Chandler's "contrastivist-error-theory" seems to presuppose).

In the final section of this paper, I will discuss "The Conjunction Fallacy." There, we will see that psychological hypotheses analogous to Chandler's "contrastivist-error-theory" hypothesis [regarding the conflation of (*) and (**) by *actual subjects*] are not borne out by the data. We'll also see another example with a similar confirmation-theoretic structure to (PIC), except, this time, it will be the logically *stronger* of two hypotheses that is confirmed$_i$ more strongly than the logically *weaker* alternative. And, this time, the examples won't be *merely* theoretical/philosophical in nature—we'll have lots of psychological data to draw upon.

5   THE CONJUNCTION FALLACY

Tversky and Kahneman (1983, p. 297) presented subjects with the (now infamous) "Linda example." In this example, subjects are presented with the following *evidence*:

(e) Linda is 31, single, outspoken and very bright. She majored in philosophy. As a student, she was deeply concerned with issues of discrimination and social justice and she also participated in antinuclear demonstrations.

Then, subjects were asked which of the following two hypotheses about Linda is *more probable, given the above evidence e about Linda.*

(h_1) Linda is a bank teller.
(h_2) Linda is a bank teller *and* an active feminist.

Most subjects answer that h_2 is more probable than h_1, given evidence e. That is, *prima facie*, most subjects report that their conditional credences are such that:

(14) $\Pr(h_2 \mid e) > \Pr(h_1 \mid e)$.

Unfortunately, (14) *contradicts the probability calculus*, according to which:

(15) If p entails q, then $\Pr(p \mid e) \leq \Pr(q \mid e)$, for *any e*.

Since Tversky and Kahneman's paper was published, there has been a great deal of ink spilled (in both cognitive science and philosophy) about what might be going on with subjects who (*prima facie*) report that their degrees of belief satisfy (14).

I won't attempt to address even a small fraction of this literature here. Rather, I will consider two possible explanations of what might be going on with such subjects that have been proposed in the recent literature. This will tie in nicely with the dialectic we've been discussing. The first explanation, which has been proposed by various cognitive scientists (Dulany and Hilton 1991; Politzer and Noveck 1991), is that when subjects are asked the question about h_1 and h_2, they are inclined to presuppose that—because h_1 and h_2 are meant to be *alternative* hypotheses about Linda—h_1 and h_2 are (or should conversationally be treated as if they are) *mutually exclusive*. To better understand this proposal, it helps to work with a finer-grained representation of the content of h_1 and h_2. To this end, I'll now work with the following notation:

(b) Linda is a bank teller.
(f) Linda is an active feminist.

With this notation in hand, we can clearly express the "contrastivist" explanation of "the conjunction fallacy" that is now on the table. The proposal is that, when subjects are asked to contrast h_1 and h_2, what they *actually* end up contrasting are the following *mutually exclusive* alternative hypotheses about Linda:

(b & ~f) Linda is a bank teller *and not* an active feminist.
(b & f) Linda is a bank teller *and* an active feminist.

Thus, or so this "contrastivist" proposal goes, when subjects report their answer to the question, they are *actually* indicating that their credences are such that:

(16) $\Pr(b \& f \mid e) > \Pr(b \& \sim f \mid e)$.

And, because (16) does *not* imply probabilistic *incoherence*, it was unfair (and premature) of Tversky and Kahneman (and others) to conclude that typical responses to the Linda question reveal any (Bayesian) *irrationality* in actual subjects.

Various experiments have been performed in recent years, which are designed to explicitly test this "contrastivist explanation" of "the conjunction fallacy." My favorite sets of experiments are reported by Tentori et al. (2004) and Bonini et al. (2004). In these experiments, subjects are asked to *bet* on the truth of various logical combinations of b and f (and similar "conjunction fallacy" conjuncts). And, *in the very same contexts*, subjects are *also* asked to perform basic *logical inferences* (for example, *conjunction elimination*) involving various logical combinations of b and f. These experiments show quite clearly that subjects—even when they are clearly presupposing that h_2 and h_1 have the *logical forms* b & f and b, respectively—tend (in proportions not too dissimilar to those seen in the original Tversky and Kahneman experiments) to *bet more money on* the truth of h_2 (b & f) than on the truth of h_1 (b). To my mind, these experiments show rather definitively that the "contrastivist" explanation of subjects' responses to the Linda question is *inadequate* (and that subjects' responses are in violation of Bayesian rational requirements after all).

In light of these recent psychological experiments involving betting and logical inference, I prefer an alternative way of "explaining" what might be going on in the Linda case. I favor "explanations" that begin by *conceding* that subjects *are* violating Bayesian rational requirements in the Linda case. What I'm more interested in is *why* subjects might tend to make the sorts of errors they make in the Linda case. That is, I'm interested in the question of whether (and to what extent) these mistakes are "understandable," from a (broadly) Bayesian point of view.

In recent joint work with Crupi and Tentori (2008), I endorse one possible way of understanding why subjects tend to report things like (14) in the Linda case. Our main idea is to begin (as Carnap cautioned us to do) by distinguishing c_f, which is just *conditional probability*, and c_i, which gauges *degree of evidential relevance*. Because (as we have already seen) confirmation$_i$ does *not* entail (SCC), this makes it *possible* (that is, *probabilistically coherent*) for subjects' credences to be such that

(16) $c_i (b \& f, e) > c_i (b, e)$

for various measures c_i of degree of confirmation as increase in firmness. That is, whereas it is impossible for e to *confirm$_f$ b $\&$ f more strongly* than e *confirms$_f$ b*, it is *not* impossible for e to *confirm$_i$ b $\&$ f more strongly* than e *confirms$_i$ b*. Our strategy is to try to identify confirmation$_i$-theoretic conditions which (i) are sufficient to entail (16), and (ii) would be accepted by most subjects (in the context of the Linda experiments). This leads to the following central confirmation$i$-theoretic result (Crupi et al. 2008):

(17) *If* the following two conditions are satisfied:
(17.1) $c_i (b, e \mid f) \leq 0$, *and*
(17.2) $c_i (f, e) > 0$

then $c_i (b \& f, e) > c_i (b, e)$, for *all* measures c_i (that have been proposed).[16]

What condition (17.1) expresses is that the claim that the evidence e about Linda is *not positively relevant* to the claim that Linda is a bank teller— *even if* it is presupposed that Linda is an active feminist. And, what condition (17.2) says is that the evidence e about Linda *is* positively evidentially relevant to the claim that Linda is an active feminist (making *no* presuppositions about Linda). It seems that conditions (17.1) and (17.2) are (intuitively) *true* in the Linda case (and, in our experience, the vast majority of subjects are inclined to agree with this assessment). Thus, it follows from (17) that (16) must also be true in the Linda case—for *all* measures of confirmation$_i$ that have been proposed in the literature. We think this robust fact about the confirms$_i$-relations in the Linda case sheds light on why subjects may be caused to give incorrect answers to questions about the confirms$_f$-relations in the Linda case.[17] As the historical literature from philosophy of science reveals, the distinction between confirms$_f$ and confirms$_i$ is rather subtle.[18] As such, it wouldn't be very surprising if people had a tendency to make mistakes in cases where the two concepts come apart (in surprising ways).

Let's return, finally, to the philosophical dialectic concerning our various accounts of favoring and contrastive/comparative confirmation. The Linda case provides a nice illustration of the fact that evidence (e) can sometimes constitute *stronger evidence* for one hypothesis (h_2) than another (h_1), *even though h_2 is logically stronger than h_1*. And, unlike "irrelevant conjunction" cases, this assessment does *not* depend on one's choice of c_i-measure. *All* Bayesian confirmation-theorists should accept that e *confirms$_i$/supports$_1$ h_2*

more strongly than h_1 in the Linda case. Moreover, the betting/logical inference experiments (Tentori et al. 2004; Bonini et al. 2004) indicate that actual subjects are *not* hearing the probability question as a "contrastive" one—*if* this requires a presupposition that the alternative hypotheses about Linda are *mutually exclusive*. This casts doubt on the Chandler-Hitchcock-style strategy (as applied to "irrelevant conjunction" cases) of substituting a contrastive question for a non-contrastive one in cases where there is a (*prima facie*) logical dependence between alternative hypotheses. In closing, I wonder whether subjects would be inclined to judge that *e favors* h_2 over h_1 in the Linda case. I conjecture that that actual subjects would *not* balk at such a claim (at least, not on grounds of non-mutual-exclusivity of the alternatives). Indeed, it seems to me quite natural to say that *e favors* h_2 over h_1 in the Linda case. Having said that, I should note that this is a burgeoning area of philosophical and psychological research. As such, much work remains to be done here—on both the philosophical and empirical sides. The good news is that this provides some exciting opportunities for future collaborative research between philosophers and cognitive scientists.[19, 20]

NOTES

1. Here, we are assuming that all bodies of evidence are *propositional*. This is a typical assumption made by Bayesian epistemologists and philosophers of science. It is not universally accepted (in epistemology generally) that all evidence is propositional. But, this is not an unpopular view either. See Williamson (2002, ch. 9) and Neta (2008) for recent discussions. Also, we're understanding conditional probability as "probability *on an indicative supposition*." This is pretty standard in the present context. See Joyce (1999) for more on this "probability on an indicative supposition" conception, and how it differs from a *subjunctive*-suppositional conception of $\Pr(\bullet \mid \bullet)$, which may be more appropriate for causal or explanatory applications. Finally, following Hajek (2003), we assume that *all* probabilities are *inherently conditional* in nature. When we talk about the "prior probability" $\Pr(p)$ of a proposition p, this is really just shorthand for a conditional probability $\Pr(p \mid K)$, relative to some unspecified background corpus K.
2. From now on, I will (for simplicity) suppress the background corpus K, unless we need to be explicit about its content. But, as I implied in note 1, all confirmation claims and quantities are (implicitly) *relativized to background corpora*. So, when we write $c(H, E)$ this is really just shorthand for $c(H, E \mid K)$, for some unspecified background corpus K (where, formally, $c(H, E \mid K)$ is obtained from $c(H, E)$ by *conditionalizing* $\Pr(\bullet)$ in $c(H, E)$ on K). The behavior of *confirmational contrasts* involving different background corpora has been underdiscussed in the literature. See Fitelson (2001) for a notable exception. I will not have the space here to discuss contrasts involving alternative background corpora. But, toward the end of the paper, we will see an example where the content of K becomes important
3. See Fitelson (2001), Joyce (2003), and Crupi et al. (2007) for contemporary discussions of the various measures of c_i that have been proposed and defended in the Bayesian confirmation-theoretic literature. My definition of c_i—as a function of the posterior and prior probabilities of H—is intentionally

restrictive (that is, intentionally less than maximally general), so as to avoid various technical subtleties that are not central to the issues I'm discussing in this article. Moreover, we take logarithms of the ratio measures to ensure that they are positive when E confirms$_i$ H (relative to K), negative when E disconfirms$_i$ H (relative to K), and zero when E is neither confirms$_i$ nor disconfirms$_i$ H (relative to K). This is merely a useful convention, which does not affect the comparative/contrastive structure imposed by the measures.

4. For the sake of brevity, I have omitted all proofs of technical claims from this article.

5. Some philosophers urge that *both* sorts of probabilities are needed. For instance, Jim Hawthorne (2005) argues that Bayesians need *both* "degrees of belief" (which are *subjective*) *and* "degrees of support" (which are *objective*, and *implied by concrete statistical hypotheses/models*).

6. There is a rather vast literature on so-called "objective Bayesianism," which tries to identify certain features of *objectively reasonable/rational* initial/prior credence functions. I won't discuss that literature here (because I don't want to dwell on issues surrounding the interpretation of confirmation-theoretic probabilities). But, "objective Bayesianism" has a rather long (and rather tortuous) history, which features many notable figures, for example, Leibniz (see Hacking (1971), Keynes (1921), Carnap (1962), de Finetti (1969), and Maher (2010); for a state-of-the art defense of "objective Bayesianism," see Jon Williamson (2010)).

7. It is no accident that the logical relations in my example are precisely the way they are. It is important that the intuitive verdict be explicable solely on the basis of *logical asymmetries* between E, H_1, and H_2 . And, this sort of structure is (more or less) the *only* one that will do the trick. See my (2007) and Chandler (2010) for further discussion of this and other putative counterexamples to (LL).

8. This also means that Hitchcock's account of contrastive probabilistic explanation (1996, 1999)—for two contrasted *explananda*, relative to a single *explanans*—*reduces to a simple comparison of the likelihoods* of the contrasted *explananda*, relative to the *explanans*. This is an interesting and important theoretical connection (and unification). It reveals that something like (LL) is presupposed in various contemporary "contrastive probabilistic explications" of *both* explanation *and* confirmation.

9. Peter Lipton (1990) voices an analogous complaint about Hitchcock-style approaches to contrastive explanation. Lipton thought that the members of explanatory contrast classes did *not* always have to be mutually exclusive. I won't be able to discuss that dialectic here. But, the parallel is worth noting. In the final section, we'll see some psychological data that support our Liptonian complaint

10. Any hypothesis H will entail/predict *many* observational consequences. It just *can't* be the case that verifying *any one* of these (myriad) predictions would be *sufficient to warrant belief in* H. It is also useful to note that confirms$_f$ does *not* satisfy (2)—not even in the case where H entails E (unlike confirms$_i$). For these reasons, I'm focusing on confirms$_i$ for the remainder of this section.

11. It follows from this strong explication of "irrelevant conjunct" that X is probabilistically independent of *all logical combinations* of H and E. That's why it's the *strongest* Bayesian explication

12. In my (2002) I adopted the strong explication of "irrelevant conjunct." In a more recent paper, Jim Hawthorne and I show that the weaker explication suffices for the Earman-style approach to (PIC) that we favor (Hawthorne and Fitelson 2004). That being said, there are certain dialectical advantages

(for a "full-blown Bayesian") to using the stronger explication of "irrelevant conjunct." See note 13 for further discussion.

13. Glymour (personal communication) criticized the "weak irrelevance" versions of our results on the grounds that our weak explication of irrelevance may classify conjuncts that are "intuitively irrelevant" as "relevant." We can avoid this worry (as well as worries about distinguishing *redundant* conjuncts and *irrelevant* conjuncts) by using the strong explication of irrelevance instead. However, this is probably not a move that a Likelihoodist would be inclined to make. See note 12 for further discussion.

14. We also think that Maher's claim about the centrality of (*) as an intuition that drives people to think there *is* a (PIC) in the first place is, at best, dubious—as a matter of *historical* fact. For instance, we seriously doubt that (*) was essential (or even *significant*) in *Glymour's* (1980) thinking about (PIC).

15. To be fair to Chandler, he didn't realize that (HC) was equivalent to (LL*) at that time (and neither did Hitchcock). That's something I pointed out to him (and Hitchcock) after his paper was published. In his more recent article, Chandler incorporates this insight into his discussion (Chandler 2010).

16. We add the parenthetical caveat "that have been proposed" in (17), because it is theoretically possible to gerrymander bizarre relevance measures that violate (17). However, none of these bizarre measures is on the table in the contemporary dialectic. For present purposes, all that really matters is that the result holds for *both* the *Likelihoodist* measure *r*, and all of the *non-Likelihoodist* measures (for example, *d* and *l*) that appear in the literature. This ensures that the Bayesian debates about the "conjunction fallacy" are (in an important sense) *orthogonal* to the Likelihoodism debate we've been discussing

17. It is worth noting that there are (equally plausible) *alternative* sets of sufficient conditions for our desired conclusion (16) that involve *only likelihoods* (relative to *e*) of different logical combinations of *b* and *f* [7, note 1]. So, even *Likelihoodists* should accept the conclusion that $\Pr(e \mid h_2) > \Pr(e \mid h_1)$—*even if* they refuse to accept the claim that *e favors* h_2 over h_1, on the "constrastivist-presuppositional" grounds that such claims presuppose that h_1 and h_2 are mutually exclusive

18. Indeed, Popper's critique (1954) of the first edition of Carnap's *Logical Foundations of Probability* (1950) was that *Carnap himself* had conflated confirms$_f$ and confirms$_i$. In the second edition of *LFP* [4], Carnap basically conceded this point to Popper. This is what led Carnap (1962, new preface) to implore his readers to be careful about distinguishing confirms$f$ and confirms$_i$ in the first place.

19. In the last few years, there has been a flurry of both philosophical and psychological work on confirmation/support judgments and their relation to probability judgments. See Tentori et al. (2007), Shogenji (2010), Atkinson et al. (2009), and Tentori et al. (2010) for more on this recent strand of philosophical and cognitive-scientific research.

20. Thanks to Fabrizio Cariani, Jake Chandler, Vincenzo Crupi, Kenny Easwaran, Hartry Field, Clark Glymour, Jim Hawthorne, Chris Hitchcock, Jim Joyce, Jim Pryor, Susanna Rinard, Teddy Seidenfeld, Elliott Sober, Dan Steel, Katya Tentori, and Mike Titelbaum for useful discussions.

REFERENCES

Atkinson, D., J. Peijnenburg, and T. Kuipers. 2009. How to Confirm the Conjunction of Disconfirmed Hypotheses. *Philosophy of Science* 76:1–21.

86   *Branden Fitelson*

Bonini, N., K. Tentori, and D. Osherson. 2004. A Different Conjunction Fallacy. *Mind & Language* 19(2):199–210.
Carnap, R. 1950. *Logical Foundations of Probability.* 1st ed. Chicago: University of Chicago Press.
Carnap, R. 1962. *Logical Foundations of Probability.* 2nd ed. Chicago: University of Chicago Press.
Chandler, J. 2007. Solving the Tacking Problem with Contrast Classes. *The British Journal for the Philosophy of Science* 58(3): 489–502.
Chandler, J. 2010. Contrastive Confirmation: Some Competing Accounts. Forthcoming in *Synthese.*
Crupi, V., B. Fitelson, and K. Tentori. 2008. Probability, Confirmation, and the Conjunction Fallacy. *Thinking & Reasoning* 14(2): 182–199.
Crupi, V., K. Tentori, and M. Gonzalez. 2007. On Bayesian Measures of Evidential Support: Theoretical and Empirical Issues*. *Philosophy of Science* 74: 229–252.
Finetti de, B. 1969. Initial Probabilities: A Prerequisite for Any Valid Induction. *Synthese* 20(1): 2–16.
Finetti de, B. 1989. Probabilism. *Erkenntnis* 31(2): 169–223.
Dulany, D., and D. Hilton. 1991. Conversational Implicature, Conscious Representation, and the Conjunction Fallacy. *Social Cognition* 9(1): 85–110.
Earman, J. 1992. *Bayes or Bust?* Cambridge, MA: MIT Press.
Fitelson, B. 2001. A Bayesian Account of Independent Evidence with Applications. *Philosophy of Science,* 68(3): 123–140.
Fitelson, B. 2001. Studies in Bayesian Confirmation Theory. PhD diss., University of Wisconsin–Madison.
Fitelson, B. 2002. Putting the Irrelevance Back into the Problem of Irrelevant Conjunction. *Philosophy of Science* 69: 611–622.
Fitelson, B. 2007. Likelihoodism, Bayesianism, and Relational Confirmation. *Synthese* 156(3): 473–489.
Forster, M. 1995. Bayes and Bust: Simplicity as a Problem for a Probabilist's Approach to Confirmation. *British Journal for the Philosophy of Science* 46: 399–424.
Glymour, C. 1980. *Theory and Evidence.* Princeton: Princeton University Press.
Good, I.J. 1971. 46656 Varieties of Bayesians. *The American Statistician* 25(5): 56–63.
Hacking. I. 1971. The Leibniz-Carnap Program for Inductive Logic. *The Journal of Philosophy* 68: 597–610.
Hájek, A. 2003. What Conditional Probability Could Not Be. *Synthese* 137(3): 273–323.
Hájek, A. 2010. Interpretations of Probability. In *The Stanford Encyclopedia of Philosophy* http://plato.stanford.edu/entries/probability-interpret/ (accessed December 21, 2011) ed. by Edward N. Zalta
Hawthorne, J. 2005. Degree-of-Belief and Degree-of-Support: Why Bayesians Need Both Notions. *Mind* 114(454): 277–320.
Hawthorne, J., and B. Fitelson. 2004. Discussion: Re-solving Irrelevant Conjunction with Probabilistic Independence. *Philosophy of Science* 71(4): 505–514.
Hempel, C. 1945. Studies in the Logic of Confirmation. *Mind* 54:1–26, 97–121.
Hitchcock, C. 1996. The Role of Contrast in Causal and Explanatory Claims. *Synthese* 107(3): 395–419.
Hitchcock, C. 1999. Contrastive Explanation and the Demons of Determinism. *The British Journal for the Philosophy of Science* 50(4): 585–612.
Joyce, J. 1999. *The Foundations of Causal Decision Theory.* Cambridge: Cambridge University Press.

Joyce, J. 2003. On the Plurality of Probabilist Measures of Evidential Relevance. Paper presented at the Bayesian Epistemology Workshop of the 26th International Wittgenstein Symposium, Kirchberg, Austria.

Keynes, J. 1921. *A Treatise on Probability*. London: Macmillan.

Lipton, P. 1990. Contrastive Explanation. *Royal Institute of Philosophy Supplement* 27: 247–266.

Maher, P. 2004. Bayesianism and Irrelevant Conjunction. *Philosophy of Science* 71: 515–520.

Maher, P. 2010. Bayesian Probability. *Synthese* 172(1): 119–127.

Milne, P. 1996. Log[$p(h/eb)/p(h/b)$] Is the One True Measure of Confirmation. *Philosophy of Science* 63:21–26.

Neta, R. 2008. What Evidence Do You Have? *The British Journal for the Philosophy of Science* 59(1): 89–119.

Politzer, G., and I. Noveck. 1991. Are Conjunction Rule Violations the Result of Conversational Rule Violations? *Journal of Psycholinguistic Research* 20(2): 83–103.

Popper, K. 1954. Degree of Confirmation. *The British Journal for the Philosophy of Science* 5: 143–149.

Ramsey, F. 1931. Truth and Probability. In *The Foundations of Mathematics and Other Logical Essays*, ed. R. Braithwaite, 156–198. London: Kegan and Paul.

Rinard, S. 2005. Comments on Dan Steel's "Bayesian Confirmation Theory and the Likelihood Principle." Presented at the Second Annual Formal Epistemology Workshop, UT–Austin, May 2005..

Rosenkrantz, R. 1994. Bayesian Confirmation: Paradise Regained. *The British Journal for the Philosophy of Science* 45: 467–476.

Royall, R. 1997. *Statistical Evidence: A Likelihood Paradigm*. London: Chapman & Hall.

Schupbach, J. 2010. Is the Conjunction Fallacy Tied to Probabilistic Confirmation? Forthcoming in *Synthese*.

Shogenji, T. 2010. The Degree of Epistemic Justification and the Conjunction Fallacy. Forthcoming in *Synthese*.

Sober, E. 1994. Contrastive Empiricism. In *From A Biological Point of View: Essays in Evolutionary Philosophy*, 114–135. Cambridge: Cambridge University Press.

Sober, E. 2006. Is Drift a Serious Alternative to Natural Selection as an Explanation of Complex Adaptive Traits? *Royal Institute of Philosophy Supplement* 80(56): 10–11.

Steel, D. 2007. Bayesian Confirmation Theory and the Likelihood Principle. *Synthese* 156(1): 53–77.

Tentori, K., N. Bonini, and D. Osherson. The Conjunction Fallacy: A Misunderstanding about Conjunction? *Cognitive Science* 28(3): 467–477.

Tentori, K., and V. Crupi. 2010. How the Conjunction Fallacy Is Tied to Probabilistic Confirmation: Some Remarks on Schupbach. Forthcoming in *Synthese*.

Tentori, K., V. Crupi, N. Bonini, and D. Osherson. 2007. Comparison of Confirmation Measures. *Cognition* 103(1): 107–119.

Tversky, A., and D. Kahneman. 1983. Extensional versus. Intuitive Reasoning: The Conjunction Fallacy in Probability Judgment. *Psychological Review* 91: 293–315.

Williamson, J. 2010. *In Defence of Objective Bayesianism*. Oxford: Oxford University Press.

Williamson. T. 2002. *Knowledge and its Limits*. Oxford: Oxford University Press.

4 Contrastive Belief

Martijn Blaauw

There is a trend in contemporary philosophy to treat many of the key philosophical concepts in contrastive terms. In *philosophy of science*, philosophers argue that the concept of "explanation" should be interpreted along contrastive lines; in *metaphysics*, philosophers argue that the concept of "causation" should be interpreted along contrastive lines; and in *epistemology*, finally, philosophers argue that the concepts of "knowledge" and "evidence" should be analyzed along contrastive lines.

I have noticed that there are two main reactions to such contrastive proposals: the *plain disbelief* and *outright excitement*. In this chapter, I will display the second reaction and try to make plausible the position that another key epistemological notion should be interpreted contrastively as well: the notion of "belief." That is, I will be arguing that belief attributions are always made against the background of a set of contrastive propositions. I will show that accepting such a "doxastic contrastivism" has a very interesting consequence: it allows us to defend a novel solution to the problem of radical skepticism.

This paper is organized in the following way. In section 1, I introduce some cases that support a contrastive reading of "to believe." In section 2, I make some preliminary remarks to delineate the topic I'm interested in. In sections 3 and 4, I will make a case in support of the conclusion that "to believe" is a contrastive notion. In section 5, I will answer an objection to the contrastive view of "to believe." And finally, in section 6, I will show in what way doxastic contrastivism can help to solve the problem of radical skepticism.

1 TEST CASES THAT SUPPORT DOXASTIC CONTRASTIVISM

In what way is doxastic contrastivism supported by our intuitions about several test cases? There are three desiderata any case must meet, due to Schaffer (2006, 88). First, we need *pairs* of cases because only pairs of cases can illustrate that changes in the contrasts can lead to a change in our doxastic intuitions regarding the cases. The cases must furthermore be *nonprejudiced*:

the cases themselves must not tell us what our intuitions should be. The cases must finally be *minimal*: they should not contain any other changes besides changes in the contrasts. So here are three such cases.[1]

Case #1

(1) Lady Victoria mutilated the Rembrandt in the museum and Holmes just found Lady Victoria's fingerprints in the exhibition room. Watson asks, "Holmes, do you believe that *Lady Victoria* mutilated the Rembrandt?"

(2) Lady Victoria mutilated the Rembrandt in the museum and Holmes just found Lady Victoria's fingerprints in the exhibition room. Watson asks, "Holmes, do you believe that Lady Victoria mutilated *the Rembrandt?*"

Case #2

(1) Lady Victoria mutilated the Rembrandt in the museum and Holmes just found Lady Victoria's fingerprints in the exhibition room. Watson asks, "Holmes, do you believe that it was Lady Victoria who mutilated the Rembrandt?"

(2) Lady Victoria mutilated the Rembrandt in the museum and Holmes just found Lady Victoria's fingerprints in the exhibition room. Watson asks, "Holmes, do you believe that it was the Rembrandt that was mutilated by Lady Victoria?"

Case #3

(1) Lady Victoria mutilated the Rembrandt in the museum and Holmes just found Lady Victoria's fingerprints in the exhibition room. Watson asks, "Holmes, do you believe that Lady Victoria rather than Professor Moriarty mutilated the Rembrandt?"

(2) Lady Victoria mutilated the Rembrandt in the museum and Holmes just found Lady Victoria's fingerprints in the exhibition room. Watson asks, "Holmes, do you believe that Lady Victoria mutilated the Rembrandt rather than the Renoir?"

It seems to me that in every case, Holmes will answer "yes" in (1) but will answer "no" in (2). That is, I have the intuition that it would be *true* to say that Holmes believes in (1), but *false* to say that Holmes believes in (2).

Note that for each case, (1) and (2) differ only with respect to stress (Case #1), cleft construction (Case #2), and rather-than clause (Case #3). Besides these changes, the cases are identical. However, intuitions reverse from a willingness to attribute belief in the (1)-cases to an unwillingness to attribute belief in the (2)-cases. Hence, it seems that we should conclude that changes in stress, cleft construction, or rather-than clause influence our doxastic intuitions. Our intuitions with respect to the pairs of cases are guided by changes in the contrasts. In the first member of each case, the contrasts are "other mutilators," whereas in the second member of each case, the contrasts are "other objects to mutilate." Moreover, the evidence Holmes possesses can eliminate that someone other than Lady Victoria mutilated the Rembrandt. But the evidence Holmes possesses cannot eliminate that something other than the Rembrandt was mutilated by Lady Victoria. Accordingly, we intuit that Holmes believes in the first member of each case, but does not believe in the second member of each case.

This provides some motivation to accept doxastic contrastivism.[2] If belief attributions are sensitive to changes in the contrasts, it seems plausible that the binary belief relation in fact has an extra argument-place Q that serves to collect the contrasts. So on doxastic contrastivism, the (1)-members of each pair should be answered by Holmes as follows:

(A)   Yes, I believe that Lady Victoria mutilated the Rembrandt rather than that Professor Moriarty mutilated the Rembrandt.

And the (2)-members of each pair should be answered by Holmes as follows:

(B)   No, I do not believe that Lady Victoria mutilated the Rembrandt rather than that Lady Victoria mutilated the Renoir.

Holmes believes that Lady Victoria mutilated the Rembrandt rather than that Professor Moriarty mutilated the Rembrandt—Holmes is an experienced detective who takes the available evidence at face value. And Holmes does not believe that Lady Victoria mutilated the Rembrandt rather than that Lady Victoria mutilated the Renoir—Holmes is a cautious detective who does not jump to risky conclusions on the basis of insufficient evidence.

2   PRELIMINARIES

But what does it mean to *believe* a proposition? If we, for instance, say that John believes that broccoli is health improving, *what do we say*? Here are two preliminary remarks.

In the first place, "to believe" apparently can take various different objects. Most often, "to believe" has a proposition as its object: John believes *that it rains*. But we also sometimes use phrases such as "S believes *in* p" (for instance "John believes in physical exercise," or "I believe in you") or "S believes R" ("John believes Jill" or "John believes his surgeon"). For present purposes, I will not address the question whether these different uses of "to believe" are reducible to one basic use, but I will focus solely on propositional belief.

In the second place, "to believe that" can be used in three different ways.[3] In order to illustrate those ways, consider the following examples:

Occurrent Belief

Jill asks John what time it is. John glances at his watch and says: "It is three o'clock." Plausibly, John believes that it is three o'clock.

Dispositional Belief

John is sleeping. Jill is on the phone with her best friend Julie. Jill says: "Well, John believes that broccoli is health improving."

Disposition to Believe

Jill is lecturing on knowledge. She says: "My husband, John, has never actively considered the proposition *that he is less than forty feet tall*, but it makes perfect sense to say that he does believe this."

These three examples illustrate three different uses of "believes." First, the use of "believes" on which we attribute that a subject, S, *occurrently* believes a proposition p. John occurrently believes that it is three o'clock: the proposition "that it is three o'clock" is currently before his mind. Second, the use of "believes" on which we attribute that S *dispositionally* believes that p. John does not occurrently believe that broccoli is health improving, yet he has occurrently believed this in the past, so it makes sense to say that he believes it now. The sense in which he does is dispositional: had the proposition that broccoli is health improving been put to John, he would have occurrently believed it. Third, the use of "believes" on which we attribute that S *has the disposition to believe that p*. John has never even considered the proposition that he is less than forty feet tall, but there is a sense in which he can be coherently said to believe this: had someone asked him whether he was less than forty feet tall or not, he would certainly have believed "that he was less than forty feet tall."

The *similarity* between dispositional beliefs and dispositions to believe is that they both involve counterfactuals: had proposition *p* been put to one's attention, then one would have occurrently believed *p*. The *difference* between dispositional beliefs and dispositions to believe is that the former category of belief involves beliefs that one has believed before, whereas the latter category of belief involves beliefs than one has not believed before. This makes that there are far more dispositions to believe than dispositional beliefs.

Putting this together, if we say "S believes that p," we might mean to say either that S *occurrently* believes that p, that S *dispositionally* believes that p, or that S *has the disposition to believe* that p. Now the key question is: which of these three types of belief is the most basic type? In what follows,

I will take occurrent beliefs to be the basic type. Dispositional beliefs and dispositions to believe are occurrent beliefs *to be*.[4]

3   BELIEF AND CONFIDENCE

So the question becomes: what does it mean to occurrently believe a proposition? If John occurrently believes that broccoli is health improving, *what does that mean?*[5]

To a first approximation, I propose that in order for S to (occurrently) believe that p, S must be *confident* that p:

> **Belief-I**: S believes that *p* means that S is confident that *p* is true.[6]

That this is a natural way to analyze the notion of belief can be seen from the fact that it sounds incoherent to say that John believes that broccoli is health improving while he is not confident that it is. If one believes that *p* is true, then one must be confident that *p* is true.

One may object that it *is* possible to believe that *p* without being confident that *p*.[7] For instance, suppose that you believe that Timothy Williamson wrote a book entitled "Knowledge and Its Limits." Further suppose someone asserts with a lot of self-assurance that Williamson wrote a book called "Unnatural Doubts." The confidence with which this is asserted makes that you are no longer confident that Williamson wrote "Knowledge and Its Limits," even though you continue to believe it. Thus, the objector concludes, you can believe that *p* without being confident that *p*. I would reply that confidence is a degree concept, and that what happens in this case is that your degree of confidence that Williamson wrote "Knowledge and Its Limits" *diminished* but did not vanish. So I resist the idea that you are no longer confident in this case.

Still, I think that **Belief-I** needs to be revised. Confidence is a contrastive concept. It is intuitive to think that there is no such thing as being confident that p *simpliciter*. One isn't confident that p *full stop*, but one is always confident that p as opposed to a set of not-p alternatives. Suppose that John is confident that his laptop is in the bedroom. This confidence only makes sense if it is contrasted with propositions such as {that it is in the living room, that it is in the bathroom}. It could well be that John isn't confident at all that his laptop is in the bedroom contrasted with the propositions {that burglars stole it, that his wife took it with her to work}. To take another example, suppose that I am confident that the meeting starts at 3:00 p.m. What does this mean? Plausibly, this means that if I am presented with some possible meeting times {3:00 p.m., 4:00 p.m., tomorrow at 3:00 p.m.}, I will pick 3:00 p.m. With respect to these alternatives, I am confident that the meeting starts at 3:00 p.m. But this does *not* mean that I will be confident that the meeting starts at 3:00 p.m. if the alternative is {that the meeting has suddenly been rescheduled to 3:15 p.m.}. What the contrasts are determines whether one has confidence in the target proposition. So the characterization of belief in terms of confidence isn't precise enough because it does not add these alternatives. *Being confident that p means being confident in a limited domain.* In what follows, I will phrase this as: S is confident that p rather than Q, where Q is the set of alternatives as opposed to which S is confident that p. Putting all this together, I submit the following updated gloss on "belief":

Belief-II: S believes that p means that S is confident that p rather than Q.

One might worry that **Belief-II** is too vague. I proposed earlier that confidence is a degree concept. But what *degree* of confidence is needed for S to believe p? Should John be *extremely* confident that the meeting starts at 3:00 p.m. rather than 4:00 p.m.? Should he be *very* confident? Would *a bit* of confidence suffice? Is *utter* confidence too strong? I doubt that there are precise degrees here, waiting to be identified. But we can sidestep this issue altogether by claiming that what is needed is simply that S is *more* confident that p than q. How much confidence that will amount to is irrelevant. So I submit:

Belief-III: S believes that p means that S is more confident that p than that Q.

In sum: the notion of confidence needed is *comparative confidence.*[8]

Here are three topics that could stand further discussion if one wants to further develop the idea of contrastive belief. First, what is the doxastic status of the contrastive propositions? Second, is the set of contrastive propositions a fixed set of propositions or a variable set of propositions? Third, is belief *itself* a contrastive concept? I now turn to answer these questions.

4 REFINEMENTS

Starting with the first question, what is the doxastic status of the propositions in Q? Suppose that John occurrently believes that it rains rather than {snows, hails, storms}, and occurrently is more confident that it rains than {snows, hails, storms}. We can now suppose that the proposition "that it rains" is before John's mind, but what about the other propositions? Here the distinction between dispositional beliefs and dispositions to believe can become useful. I propose that the doxastic status of the propositions in Q is as follows: if they had been put to John's attention, he would have been less confident that they were true than that p is true. So if the propositions "that it snows," "that it storms," and "that it hails" had been put to John's attention, he would have been less confident that they were true than that it rains. Conversely, he would be more confident that it rains rather than that they were true.

Moving on to the second question, suppose that we hold that confidences should be understood against the background of a set of contrastive propositions. Which propositions will go in the set of contrastive propositions? All contrastive propositions, or just a subset? I would say that the content of the set of contrastive propositions is at least in part determined by what concepts the subject has actually available. Suppose that John has never heard of sleet and does not have this concept available. He has heard of snow, however, and does have this concept available. Further suppose that it is raining. In that case, John will be confident that it is raining rather than snowing. But he will not be confident that it is raining rather than sleeting because he doesn't have the latter concept available. One's confidences are constrained by which concepts one has available.

This answer might raise the following worry. Suppose that John *is* confident that it rains as opposed to that it {snows, hails, rains ice cream, rains great pumpkins, and so on}. This sounds quite suspicious. On the one hand, we would want to say that John *is* confident that it rains rather than all those possibilities. He has the relevant concepts available. On the other hand, it somehow seems too broad and far too inclusive. We could go on indefinitely. How to solve this problem?

I propose to solve it by making a distinction between *the state of being more confident that p rather than Q1* and *attributing that S is more confident that p rather than Q1*. Although the state of being more confident that p rather than Q1 is fixed and includes all propositions in Q1, *attributing* that S is more confident that p rather than Q1 makes the content of Q1 variable. So John is in a state of being more confident that it rains rather than {snows, hails, rains ice cream, rains great pumpkins}. Yet if Jill attributes that John believes that it rains, then she attributes that John is more confident that it rains rather than, for instance, {snows, hails, storms}. Belief *attributions* always concern a variable subset of Q1, and this explains our reluctance to say that John is confident that it rains as opposed to a very large set of contrastive propositions.[9] So the suspicion is due to the fact that it sounds strange to attribute belief that *p* as opposed to a very broad contrast class. Yet that does not mean that the state of being confident that p isn't opposed to a very broad contrast class.

Arriving at the third question, suppose that we interpret belief as expressing comparative confidences as per **Belief-III**. In that case, shouldn't we also interpret belief itself in contrastive terms? Put differently, if "to believe" is to be analyzed in terms of comparative confidence, then shouldn't "to believe" *itself* also display the contrasts that comparative confidence displays: S believes that p rather than Q? I think an argument can be made for this. For now, let me simply point out that it sounds intuitive to uphold an account of contrastive belief based on the fact that "to believe" means "to be more confident that *p* rather than Q." One might, of course, object that arguing for a contrastive account of belief on this basis would be to perform a fallacy of composition. For instance, the facts that knowledge implies justification and that justification comes in degrees cannot imply that knowledge itself comes in degrees. Likewise, the facts that belief implies comparative confidence and that comparative confidence is contrastive cannot imply that belief itself is contrastive. I would reply that I don't argue that belief that *p implies* comparative confidence that p; I argue that to believe p simply means to *have* comparative confidence in *p*.

5 CONTRASTIVE BELIEF AND TRUTH CONDITIONS

I now turn to another question that should be addressed when developing an account of contrastive belief. What are the *truth conditions* for contrastive belief?[10] In this section, I will argue that two options to specify the truth conditions of contrastive belief can be defended from objections.

If one accepts the standard view that "to believe" simply expresses a two-place relation between a subject and a proposition, then the truth conditions for belief can be specified as follows:

Binary Belief and Truth: The belief that p is true iff p.

On this view, S's belief that it rains is true if and only if it rains. But what would an account of the truth conditions of belief look like if we accepted that belief is contrastive? One option, suggested by Baumann (2008), would be straightforward:

Contrastive Belief and Truth 1: The belief that p rather than q is true iff p rather than q.

On this view, S's belief that it rains rather than snows is true if and only if it rains rather than snows. Baumann (2008, 198) argues that the problem with this option is that it is difficult to understand what kind of condition it specifies: "We don't understand what kind of condition that is and what the right-hand side of the conditional means. What does it mean to say that 'p rather than q'?" In response to Baumann's worry, I am not so sure that it really is that difficult to understand what "p rather than q" means. Imagine that John says "I believe that it rains rather than snows." And suppose that you wonder—but does John have a true belief? Given that it in fact rains—and doesn't snow—it seems natural enough to conclude that John has a true belief. Imagine that John says "I believe that it rains rather than sleets." And suppose that you wonder—but does John have a true belief? Given that rain and sleet can sometimes be hard to distinguish, in this case it might be difficult to tell. But this isn't due to the fact that belief is contrastive. Indeed, the contrast *helps to specify why it is in this case difficult to tell whether John has a true belief*: rain and sleet are sometimes hard to distinguish. A binary account of belief wouldn't have this way of specifying why it is hard to tell whether John has a true belief. So thus far, I'm not convinced by Baumann's objection.

Another way to specify the truth conditions of contrastive belief suggested by Baumann would be:

> **Contrastive Belief and Truth 2:** The belief that p rather than q is true iff p.

But Baumann thinks this specification leads to the following problem: the belief that p rather than q would have the same truth conditions as the belief that p rather than r (Baumann 2008, 199). So John's belief that it rains rather than snows would have the same truth conditions as his belief that it rains rather than sleets, namely "that it rains." But how could these two, substantially different, beliefs have the same truth conditions, Baumann worries (2008, 199)? Note that the assumption underlying his worry is that if there is a substantial difference between two beliefs, this should be reflected in the truth conditions of these beliefs.

In response, even though I agree that the belief that p rather than q is substantially different from the belief that p rather than r, I think we can—and should—resist accepting that this should be reflected in the truth conditions of these beliefs. It might help here to compare belief and *conjecture* to see how we could resist Baumann's assumption.[11] Belief that p and conjecture that p have the same truth conditions. For if someone has a true belief that p her conjecture that p will be true as well:

> The belief that p is true iff p.
> The conjecture that p is true iff p.

But, crucially, believing that p and conjecturing that p are two substantially different states. You can conjecture truly but not believe truly. So even though belief and conjecture are different states—just like the belief that p rather than q is a state substantially different from the belief that p rather than r—they can be coherently thought of as having the same truth condition. So I think there is good reason to resist Baumann's assumption.

In conclusion, there are at least two ways to understand the truth conditions of contrastive belief. My aim here isn't to determine which of them is correct. My only aim is to neutralize the objection that a coherent account of the truth conditions of contrastive belief couldn't be given.

6   RADICAL SKEPTICISM: CLOSURE AND UNDERDETERMINATION

In this section, I will show what the puzzle-solving potential of an account of contrastive belief is. Specifically, I will show how an account of contrastive belief can help to solve the problem of radical skepticism.

Recent years have seen a surprising revival of interest in the problem of radical skepticism. The version of radical skepticism that most epistemologists try to respond to has the following familiar form (where "SH" stands for a skeptical hypothesis—such as that we are currently a brain in a vat—and where "O" stands for an ordinary proposition about the external world—such as that we have hands):

Skeptical Argument

(1) S does not know that not-SH.
(2) If S does not know that not-SH, then S does not know O
(3) Therefore, S does not know O.[12]

Although not many people will agree with the conclusion of this argument, it has proven surprisingly hard to say anything against it. The first premise of the argument certainly is very compelling: how could we ever eliminate the possibility that we are brains in vats, if the brain-in-a-vat hypothesis is set up in such a way that it is *beyond* elimination? The second premise of the argument is very compelling as well and is supported by the highly intuitive closure principle for knowledge. The closure principle for knowledge says, roughly, that if one knows a proposition p and if one also knows that p implies q, then one must also know q.[13] Thus, if S knows that she has hands, and if S knows that if she has hands, she is not a brain in a vat, then S also knows that she is not a brain in a vat. Finally, the conclusion of the argument is highly implausible: *of course* we know propositions about the external world.

Now skeptical arguments typically show something about the relationship between evidence, belief, and knowledge. In particular, they show that the evidence we have for beliefs about the external world can never be such that those beliefs are instances of knowledge. So suppose that I believe that I have hands. What the skeptic shows is that the evidence I have in favor of this belief can never be such that it turns the belief into an instance of knowledge. The reason is that the evidence I have in favor of this belief is simply my perception of hands. But crucially, I would have had this evidence even if I were a brain in a vat. The evidence of seeing hands does not uniquely support the belief that we have hands—it also supports the belief that we are being deceived into believing (falsely) that we have hands. The evidence is compatible with both scenarios.

Recent responses to the skeptical argument have been of three different types. The first type of response (neo-Mooreanism) denies the first premise of the argument and argues that we can know that we are not brains in vats.[14] The second type of response (anti-closure) denies the second premise of the argument by denying the closure principle which supports it.[15] And the third type of response (contextualism) defends a position according to which the skeptical conclusion is true in (skeptical) contexts where demanding standards for knowledge are in play, but false in (everyday) contexts where undemanding standards for knowledge are in play.[16]

All three answers focus on *evidence*. The neo-Moorean response focuses on evidence in that it is argued that we *can* have sufficient evidence in favor of the belief that we are not brains in vats, provided that the world is as we think it is. That is to say, assuming that skeptical worlds are far-off in logical space, we can, on the neo-Moorean view, know propositions about the external world. The anti-closure response focuses on evidence in that it is argued that although we can never have sufficient evidence in favor of the belief that we are not brains in vats, we can have sufficient evidence for beliefs about the external world nonetheless. The contextualist, finally, focuses on evidence in that it is argued that we will have sufficient evidence for the belief that we are not brains in vats in contexts in which low standards for knowledge are in play, although we will not have sufficient evidence for those beliefs in contexts in which high standards for knowledge are in play.

All three types of answer are problematic. The contrastivist about belief has the resources to answer the skeptical challenge in another way. The contrastivist about belief no longer focuses on evidence, but focuses on the *beliefs* the evidence is supposed to support. Thus the skeptic will be answered by saying that epistemologists have the wrong view on the adicity of the belief relation. *That* is what causes all the trouble. So suppose that John believes that he has hands rather than stumps. Is this particular belief an instance of knowledge? The doxastic contrastivist can answer "yes," which is the intuitively correct answer. The evidence John has in favor of this belief—his perceiving hands—does in fact support it. It supports that John has hands and it eliminates that John has stumps. Thus it seems plausible to say that John *knows* that he has hands rather than stumps. Now suppose that John believes that he has hands rather than brain-in-a-vat images of hands. Is this particular belief an instance of knowledge? The doxastic contrastivist can answer "no," which is the intuitively correct result. The evidence John has in favor of this belief—his perceiving hands—does not support it. And the reason is that although it supports that John has hands, it does not eliminate the alternative that John is perceiving mere brain-in-a-vat images of hands. Thus, it does *not* seem plausible to say that John knows that he has hands rather than brain-in-a-vat images of hands.[17]

This, in essence, is the contrastivist response to radical skepticism. Think of the relationship between evidence and belief as if the evidence were beams supporting a (doxastic) roof. According to the skeptic, the beams are too weak to support the roof. The traditional answers are to either strengthen the beams or show that they are not as weak as the skeptic thinks they are. The answer I propose is to keep the beams exactly as they are but *make the roof less heavy*.[18]

NOTES

1. Also see Schaffer (2008) for similar cases in defense of contrastive knowledge.
2. Interestingly, then, similar cases seem to support both contrastive knowledge (Schaffer 2008) and contrastive belief.
3. See for an early statement of the distinction between occurent belief and dispositional belief Gilbert Ryle (1949). Robert Audi's (1994) distinguishes between dispositional belief and disposition to believe.
4. Of course, one can argue about the criteria one uses to determine which category of belief is the basic one. Here, I use the criterion that dispositional beliefs strive toward occurrent beliefs. There is no such thing as a dispositional belief that can never become occurrent. Because dispositional beliefs are occurrent beliefs to be, occurrent beliefs are the basic category.
5. One might hold that the three types of belief have radically different analyses. However, I find this implausible, if only because we never explicitly say that someone has an occurrent belief, dispositional belief, or disposition to believe. We do say that someone believes something—and this belief might be of any one of the three types.
6. I am grateful to discussion with Jonathan Schaffer here.
7. Thanks to Duncan Pritchard for presenting this kind of objection.
8. Richard Swinburne defends a similar view in his (2001). For instance, he writes: "Belief, I suggest, is a contrastive notion; one believes this proposition as against that proposition" (34).
9. A further worry is that some contrastive propositions just seem outrageously bizarre. Who ever considers that ice cream might rain from the sky? But the fact that these propositions are bizarre does not imply that one cannot be confident that they are not the case. Indeed, I would say that the more bizarre a proposition is, the more confident we are that it is false.
10. I am grateful to Peter Baumann, Adam Morton and Walter Sinnott-Armstrong for discussion and correspondence.
11. Adam Morton has proposed this particular answer to Baumann's objection to me.
12. Versions of this argument figure in DeRose (1995), Sosa (1999), Cohen (2000), Pritchard (2002a), and Schaffer (2005).
13. For more on the closure principle for knowledge, see the discussion in Hawthorne (2004).
14. This type of response has been defended by Sosa (1999), Pritchard (2002b), and Black (2002), for instance.
15. Dretske (1970) and Nozick (1981) have defended this position.
16. For some of the key contextualist papers see DeRose (1995, 1999), Lewis (1996), and Cohen (2000).
17. For more on contrastive knowledge, see the chapter by Adam Morton in this volume. Also see Schaffer (2005, 2006) and Blaauw (2008a, 2008b, 2009). For an elucidation of how contrastive belief and contrastive knowledge relate, see Blaauw (manuscript).
18. I am grateful to audiences at Dartmouth College, the University of Aarhus, the University of Aberdeen, the Institut Jean Nicod, the University of Copenhagen, Delft University of Technology, and the University of Southern Denmark for useful comments and questions. I am particularly grateful to Peter Baumann, Julia Driver, Robert Fogelin, Klemens Kappel, Adam Morton, Duncan Pritchard, Adina Roskies, Jonathan Schaffer, Walter Sinnott-Armstrong, Jason Stanley, Roy Sørensen, and Rene van Woudenberg for feedback on earlier versions of this paper. Work in this area has been made possible by an "Overseas Conference Grant" of the

British Academy and a NWO-VENI research grant of the Netherlands Organization for Scientific Research (NWO).

REFERENCES

Audi, R. 1994. Dispositional Beliefs and Dispositions to Believe. *Noûs* 28: 419–434.
Baumann, P. 2008. Contrastivism rather than Something Else? On the Limits of Epistemic Contrastivism. *Erkenntnis* 69: 189–200.
Blaauw, M. 2008a. Contesting Pyrrhonian Contrastivism. *Philosophical Quarterly* 58(232): 471–477.
Blaauw, M. 2008b. Contrastivism in Epistemology.*Social Epistemology* 22(3): 227–234.
Blaauw, M. 2009. Contra Contrastivism. *Philosophical Issues* 18: 20–32.
Blaauw, M. *Manuscript. Knowledge in Contrast*. Book manuscript.
Black, T. 2002. A Moorean Response to Brain-in-a-Vat Scepticism. *Australasian Journal of Philosophy* 80: 148–163.
Cohen, S. 2000. Contextualism and Skepticism. *Philosophical Issues* 10: 94–107.
DeRose, K. 1995. Solving the Skeptical Problem. *Philosophical Review* 104: 1–52.
DeRose, K. 1999. Contextualism: An Explanation and Defense. In *The Blackwell Guide to Epistemology*, ed. J. Greco and E. Sosa, 187–205. Oxford: Blackwell.
Dretske, F. 1970. Epistemic Operators. *Journal of Philosophy* 67: 1007–1022.
Hawthorne, J. 2004. *Knowledge and Lotteries*. Oxford: Oxford University Press.
Lewis, D. 1996. Elusive Knowledge. *Australasian Journal of Philosophy* 74: 549–567.
Nozick, R. 1981. *Philosophical Explanations*. Oxford: Oxford University Press.
Pritchard, D. 2002a. Recent Work on Radical Skepticism. *American Philosophical Quarterly* 39: 215–257.
Pritchard, D. 2002b. Resurrecting the Moorean Response to the Sceptic. *International Journal of Philosophical Studies* 10: 283–307.
Ryle, G. 1949. *The Concept of Mind*. London: Hutchinson.
Schaffer, J. 2005. Contrastive Knowledge. In *Oxford Studies in Epistemology*, vol. 1, ed. T.S. Gendler and J. Hawthorne, 235–271. Oxford: Oxford University Press.
Schaffer, J. 2006. The Irrelevance of the Subject: Against Subject Sensitive Invariantism. *Philosophical Studies* 127: 87–107.
Schaffer, J. 2008. The Contrast-sensitivity of Knowledge Ascriptions. *Social Epistemology* 22, 235–245.
Sosa, E. 1999. How to Defeat Opposition to Moore. *Philosophical Perspectives* 13: 141–154.
Stanley, J. 2005. *Knowledge and Practical Interests*. Oxford: Oxford University Press. Swinburne, R. 2001. *Epistemic Justification*. Oxford: Clarendon.

5   Contrastive Knowledge

Adam Morton

The concept of knowledge is a very ordinary one, in spite of its philosophical glory. Like the concepts of a thing or a person or an animal, or the concepts of cause, or of action, it is one that we use every day and would be lost without. We use it when we explain people's actions ("She visited you because she knew you would never visit her"), and when we say what testimony should be trusted ("She knows where he hid the loot: I'd pay attention to what she says and does"), and when we justify our own actions ("I looked in the fridge because I didn't know where else it might be.") This ordinariness makes philosophical skepticism threatening in an immediate way. But it combines uncomfortably with the high intellectual demands of some philosophical accounts of knowledge. There are reasons for these high demands arising from the function the concept has historically played in philosophy, associating it with the ideal of a rational intellectual agent. To reconcile the pulls from everyday life and the pulls from philosophy we need to understand the reasons why we have the concept: what are the core functions that it serves in our ordinary thinking?

The claim of this paper is that the everyday functions of knowledge make most sense if we see knowledge as contrastive. That is, we can best understand how the concept does what it does by thinking in terms of a relation "a knows that p rather than q." There is always a contrast with an alternative. Contrastive interpretations of knowledge, and objections to them, have become fairly common in recent philosophy. The version being defended here is fairly mild in that there is no suggestion that we cannot think in terms of a simpler not explicitly contrastive relation "a knows that p." Some, for instance Schaffer (2005b) and Karjalainen and Morton (2003), have hinted that this stronger possibility may be right. But all that I am arguing now is that facts that are easily expressed in contrastive terms are vital to understanding why we need the concept of knowledge. In a piece that is in some ways a companion to this one (Morton 2010), I give a general survey of theories of contrastive knowledge and the differences between them.

1   BENEATH PROPOSITIONAL ATTITUDES: TRACKING

Knowledge is a factive relation: it holds between people and actual facts. You cannot know something that is not so. Facts are problematic things, somewhere between situations and propositions. Most relations between people and things, as between things and things, are just that, holding between individuals without involving anything sentence-like. And they hold just between the individuals they hold between, and not between nearby or alternative individuals. So when using them to explain or predict, some explicit or implicit reference to laws of nature, causation, or counterfactuals—something in the realm of the nomic—is needed. Many basic epistemic relations connect individual people to individuals in their environment: *a* sees o, *a* perceives o, *a* recognizes o, *a* remembers o. Sometimes we use an embedded sentence to describe what is really a relation between individuals: *a* knows that o is at location l (*a* locates o at l), *a* knows which species o belongs to (*a* classifies o). I shall assume that one important and central function of noting and stating such relations is to help anticipate and explain actions of individual people directed at individuals. Why did Alfred duck? Because he saw the stone whizzing toward his head. Why did Agatha return to the crossroads? Because she located her cell phone there. It takes some care to formulate these without using propositional attitude terms, and I take this to be due to the way that propositional attitude language dominates our descriptions of sentient life. The point, however, is that explanations couched in this language can succeed because of the multitude of relations by means of which people can direct their actions at objects. (It would be easier to describe this if I could assume that propositional attitudes are a linguistic veneer over an underlying pattern of relational thought. But although I think something of this sort may be true, and have begun to explore the idea elsewhere—see Morton (2009)—this is not the occasion to defend it.)

Note the way that the epistemic element in these explanations ("saw," "located"), serves both as a fact and as a law. In saying that Alfred saw the stone (and its trajectory) we are saying that Alfred's information state is related in a particular way to the stone and what it was doing. There is a flow of information from the stone to Alfred, a flow being a causal process that relates objects in one kind of situation to a characteristic kind of result. Combined with tacit assumptions about his tendency to avoid injury, it is as if there is a causal flow from the motion of the stone to Alfred's action. It is a flow that we the observers or describers can ride along, as when we see him duck and duck ourselves in order to avoid the projectile we take him to be avoiding. The "width" of the flow is left vague in the explanation, so we are not told exactly what other things Alfred did or would have been able to see. It may be necessary to say more about this, as we do when explaining why Alfred avoided the stone thrown by Martha but not the tomato thrown by Nelly. "There was a branch in the way, so that when he turned his head the tomato was behind it. If Martha had thrown the stone up rather than straight he wouldn't have seen it either." (I am appealing to what I take to be common to most accounts of explanation, although my expression may seem Hempelian. The idea of a flow of information is Dretske's. See Dretske 1981).

We direct our actions at objects by keeping track of them. We note their locations and attributes and how they change. Central to this are tracking relations: our representations of things are causally linked to the locations, colors, and other attributes of objects, so that if these change so do our representations. A cat chasing a mouse leaps where the mouse is. These require a certain sensitivity to the state of the object, which is supplied by different mechanisms for different aspects of the state. They all feed into a relatively uniform way of anticipating the actions of human and nonhuman agents, however: they predict that some aspect of a relation between the agent and an object will remain invariant under changes of surrounding conditions. The cat's direction vector will remain pointed at the mouse; the bird-watcher's thumb will approach the page of the book for that kind of bird. The central point for present purposes is that in mere mortals these mechanisms of sensitivity are of limited accuracy and scope. We can keep track of where a prey animal is as long as it has not taken certain evasive actions (and a predator can keep track of us as long as we have not taken certain measures). We can keep track of roughly where it is, close enough for most of our purposes but not for all conceivable purposes. Even when normal conditions of detection are satisfied, we may be unable to distinguish between usual and rare trajectories or attributes: if the mouse is between the cat and the sun, the cat may be misled by its reflection in a stream. As a result, tracking is inevitably contrastive. The cat can locate the mouse as being in front of it rather than ten degrees to the left, but not as being in front of it rather than one-half degree to the left. And not as being in front of it rather than between it and the setting sun, above the reflective stream.

Contrastivity is inherent in tracking, and tracking is basic to the purposes for which we use attributions of knowledge. I have just suggested how tracking connects with explanation and prediction, although obviously there is a lot more to say. Tracking also has a natural connection with testimony, via behavior that indicates an agent's link to an object. Suppose that we are using a dog to track some prey. After sniffing around, the dog sets off in a definite direction. We take the dog to be tracking the prey, literally, and follow her. We follow her because we take there to be a counterfactual link between her behavior and the path taken by the prey. If the prey had gone a different way the dog would have set off in a different direction, so her behavior "tells" us which way to follow it. Here too there are limits. Dogs are notoriously prone to following a trail in the wrong direction. So we are told that the prey has gone along this route, but are less sure that it is toward the end rather than the beginning.

Tracking, with its limits, connects to the aspirations of inquiry, too. An owner training a hunting dog wants it actually to track. On any particular occasion the dog's setting off in the direction of the prey does not count as success unless the dog would have gone off in another direction if the prey had taken another direction. Within limits: one doesn't expect a well-trained dog to fly, if the prey has taken a ride on a helicopter. A person wanting to become a good bird-watcher aspires to saying "finch" when it is a finch, and wants that had it been a grosbeak instead she would have said "grosbeak." She does not aspire to telling two-year-old finches from twenty-five-month-old finches. An apprentice astronomer who guesses which planet is near the horizon is rebuked by his mentor even if he has guessed right: if it had not been that planet he would have made the same guess.

I have avoided the verb "know" throughout this section. (And I have only said "knowledge" once.) That is the point. The focus is on the purposes for which we use the concept of knowledge, and how they inevitably bring in considerations of the aspects of an object that an agent is and is not in contact with, themselves naturally expressed in a "rather than" idiom. When the topic is human belief we express many of these ideas in terms of knowledge, and then to express the contrastivity we say "knows that p rather than q." But the roots lie deeper, and even if one balks at ascribing knowledge to, say, tigers, one will need to be able to say that, for instance, the tiger has traced one to one's hiding place, that this means that if one had hidden in the next bush it would have found one there, and that it does not mean that if one had taken a ride out of the park it would have been waiting at the hotel. The tiger has located one at this bush rather than that one, but not at this bush rather than the hotel.

2   CONTRASTIVE COUNTERFACTUALS

In an important article on contrastive causation, Jonathan Schaffer, who has played a central role in making philosophers take the idea of contrastive knowledge seriously, introduced the idea of a contrastive counterfactual (Schaffer 2005a). If c rather than c' had occurred, then e rather than e' would have occurred. In the way of understanding this that I think is most relevant to our current concerns here, c' and e' are actual events. It is possible that there is a workable definition of this conditional in terms of the Lewis-Stalnaker counterfactual, or in terms of more recent refinements of it. (A question that deserves serious attention is how it relates to the non-contrastive conditional.) It is also possible that it is a more fundamental idea, and the order of explanation ought to go in the other direction. This possibility becomes more plausible when we make the connection with contrastive knowledge, in particular with tracking.

We are fairly comfortable ascribing knowledge to individuals in many cases when they perceive their environment, even when the perception could easily have been fallible. In such cases, tracking analyses can give the intuitively wrong answer. For example: a person is watching a fly cross in front of a window. The window is partially reflective and a moment later a different fly will pass behind the person so that its image will appear on the glass, making it seem as if a fly is taking the path of the first fly. The person's visual systems are working well, and she comes to think that there is a fly directly in front of her, as there is. The flight of both flies is very erratic, however, so that it could easily have been that the second fly was the one she saw, which would have been behind her. If the details are spelled out suitably, it is natural to say that she knows that fly one is directly in front of her (although we might conclude that it is natural but wrong, if we had good enough reasons). One consideration we might use to back up the ascription is that she was tracking the fly: if it had been a little higher or a little to the left she would have taken it to be higher or to the left. But it is not clearly true that if the fly had not been directly in front of her she would not have taken it to be directly in front of her. Which is the more likely ("nearer," "more accessible") possibility: the possibility in which fly one is not there because it is further away and she sees it as further away, or the possibility in which fly two is behind her and is taken to be fly one in front of her? The balance between these may be very delicate, and the English conditional is surely imprecise enough that there is sometimes no answer. But the more focused contrastive conditional "If fly one had been one degree to the left of center, she would have seen it one degree to the left of center," may be unproblematically true.

The essential point is that a simple counterfactual "if p then q" may lack a truth value, or have one that is extremely hard to determine, whereas a contrastive counterfactual "if p rather than p' then q rather than q'" is straightforward. And it is very plausible that very often when we take a conditional as true we implicitly supply a pair of contrasts, which contrasts depending on context. Our evaluation of "If she had said that to me, I would have been insulted," may be different depending on whether we take it as "if she had said that rather than . . .," "if she had said that to me rather than to you . . .," ". . . I would have been insulted rather than amused," ". . . it would have been me rather than you who was insulted," and various combinations of these. So too apparently simple attributions of knowledge change their plausibility when we highlight different alternatives. She knew that fly one was *there* then; she knew that fly one was there *then*; she knew that fly *one* was there then; she knew that *fly* one was there then. And very often these contrasts will correlate with switches from one contrastive conditional to another, indicating one tracking relation rather than another. To say this is not to present an analysis of knowledge, contrastive or otherwise, in terms of tracking. The suggestion is just that considerations about tracking can influence our judgments about what a person knows, and that tracking, being a counterfactual concept, is sensitive to the contrastive considerations that tune our judgments about counterfactuals.

These considerations connect with observations made by both critics and defenders of tracking analyses of knowledge, to the effect that when defending the idea that to know one must track, we are choosy about what we are to count as a near alternative to the actual situation. A forceful exposition is found in Sherrilyn Roush (2005). (I am thinking especially of Chapter 4; I do not mean to endorse the details of what Roush is describing, which may well be right but which I find very hard to follow.) For example, if someone looks up and knows by seeing that it is a full moon, we do not consider the nearest situation in which at that very moment the moon is not full. That may be a situation in which the moon does not exist or the processes that form it give it a different orbit. In that situation that very person may not exist. Rather, we think in terms of "if she had been standing there on a day on which the moon was not full . . .," or "if she had been somewhere from which the moon was not visible," or "if Venus had been shining near that spot in the sky." But these are different, and correspond to "knows that the moon is full rather than in some other phase," "knows that it is the moon that is presenting that full appearance," "knows that it is a full moon rather than a reflection in a contact lens." (Of course we can also attribute "knows that it is a full moon rather than an alien visitation," and even "knows that the moon is full rather than her medications producing hallucinations"; see the discussion of full contrasts in section 5.) The conclusion to draw is that we describe the information-management of humans and other creatures in terms of how they keep track of facts around them, that these are very sensitive to the contrasts we read into associated conditionals, and that we take such informational states into account in attributing knowledge, which thus acquires an often hidden contrastivity.

3   EVIDENCE

Processes of the kind that allow us to keep track of things, although fundamental, are just one source of knowledge. Another basic source is the force of evidence. We do not have a generally accepted understanding of the relation between evidence and theory in the philosophy of science, or indeed in statistics. But many cases are uncontroversial, and some general facts are fairly well established. One is the essential role of background assumptions, especially those that determine the probability that some observable consequence will be found if a hypothesis is true and if its main rivals are true, and which determine which hypotheses are the main contenders. Another is the role of those alternative hypotheses, leading to the statistical dialectic of null hypothesis, alternative hypothesis, and test of significance.

Both of these facts lead to contrastivity in the force of evidence. Suppose that we want to know whether a coin is fair. We assume that it has a constant bias to heads or tails, which will be zero if it is fair. We toss the coin twelve times, and observe that it falls HHTTTHTHTHTT, five Heads and seven Tails. Calculating, we find that this would be very unlikely if the coin had a strong bias to H (for example such that it will on average land heads seven-tenths of the time), and fairly probable if the coin is fair. So the null hypothesis of Fair is favored over the alternative hypothesis of strong bias to heads. But our experiment has told us nothing to rule out a different hypothesis, that it has a much lesser bias to heads. And it does not provide evidence against the hypothesis that the probability of heads varies from one toss to another. (Perhaps as the coin ages the distribution of metal in it alters; after all the ratio in the first six tosses was 3:3, and in the next six was 2:4.) One might conclude from this in the vocabulary of Fisherian statistics, that we now know that the extreme-H hypothesis can be rejected, but not that the null hypothesis is true. Yet this leaves out the possibility that from this or possibly more similar evidence we can become reasonably certain that the coin is fair. Bayesians will stress this point, although they too are uncomfortable talking of knowledge. What does seem clear is that one can have good evidence to decide between Fair and Very Biased, inadequate evidence to decide between Fair and Slightly Biased, and no evidence at all to decide between any of these and Variable Probability. Suppose that the coin is fair, and that the fact that it is fair rather than biased is a cause of its exhibiting the kind of pattern of which HHTTTHTHTHTT is an instance. (This second fact could be seen as reducing to the truth of suitable contrastive counterfactuals.) Then one can be said to know that it is fair rather than very biased, and not to know either that it is fair rather than slightly biased or fair rather than having no constant bias.

These are not essentially different from contrastive knowledge based on limited powers of discrimination. Suppose you can tell dogs from cats but not from wolves, and you correctly identify the animal before you as a dog rather than a cat, but should not be treated as a good source on whether it is a wolf. You must be using some clues about what distinguishes dogs from cats. They may be obvious clues, but they may also be subtle and hard to access consciously, as might be if you can tell small dogs from large cats in the moonlight. These serve the role of evidence: the characteristic dog walking gait is like the run of five heads and seven tails. Or to put it differently, your ability to use either a string of random data or a typical dog feature, in order to evaluate the suggestion that you are dealing with a dog or a fair coin, is a limited discriminatory ability, which can indicate that some possibilities rather than others are actual.

It is worth stating explicitly here that the ability to discriminate two possibilities does not establish contrastive knowledge, if it is taken to mean just that if A or B is the case then it must be A. If your most informative report is "The coin is more likely to be fair than very biased, although it might be slightly biased," or "if it's either, it's a dog," then what you have is not contrastive knowledge, at any rate not of the species or of the bias. In order to know that it is a dog rather than a cat, you must first believe that it is a dog, and then your belief must be linked in an appropriate manner to the fact that it is a dog. What counts as an appropriate manner is something that divides epistemologists, in particular internalists and externalists, for reasons that thinking contrastively is not going to dissipate. But many cases are uncontroversial, and it is clear that evidence is often essential, that a discriminatory skill is often essential, and that both typically separate one possibility from others, leaving further possibilities uneliminated. The requirement of evidence or a causal connection comes in here. In some cases the evidence that A is more likely than B has to be accompanied by reasons for ruling C out as a possibility. These reasons may consist not in direct evidence but in general considerations deriving from the structure of one's system of beliefs. In other cases the discriminatory capacity that tells As from Bs has to be accompanied by facts that make Cs rarely occur in the circumstances. In yet other cases these factors will be combined. One has evidence that supports the null hypothesis A in contrast to the alternative B, but does not eliminate alternative C, but in the circumstances of enquiry C is not to be found except when something really weird is going on. And in some cases C will be an unevidenced but not unreasonable assumption, as described in the next section. It is hard to say which of these is the more fundamental element. Deep issues in epistemology arise here.

4   ASSUMPTIONS AND CONSEQUENCES

Seeking evidence that a coin is fair, you assume it has a constant bias. Trying to tell whether an animal is a dog or a cat, you assume that it belongs to one of your neighbors. In neither case do you have anything like direct evidence, but in both cases, let us suppose, your assumption is a sensible one. What makes it sensible is a controversial matter, as suggested just above. Having made the assumption, you use it in the formation of further beliefs, typically in eliminating alternatives to allow available evidence to get a grip on a situation or in allowing limited discriminatory capacities to operate effectively. You then treat some of these further beliefs as if they were definitely established. You take yourself to know them, in spite of the element of stipulation in their history.

Many epistemological theories will find this troubling. How can knowledge be based on mere assumption, even sane assumptions that in fact are true? From a contrastive point of view the situation is more manageable. You know that it is a dog rather than a cat, although you do not know that it is not a raccoon that has wandered far from its usual habitat. You know that the coin is fair rather than strongly biased to heads, although you do not know that it is fair rather than of varying bias. This does not mean that beliefs downstream from any arbitrary assumption which happens to be true can count as knowledge. At the very least it has to be an assumption that is not undermined by other things you know and believe, and it has to be an assumption that you need to make in order for your enquiry to proceed. And at the very least the epistemic grounds for discriminating between the possibilities that, armed with the assumption, you can separate, have to be solid. Of course an illuminating account of when beliefs based on an assumption are known would be extremely valuable.

But deciding what account of these matters is right cannot be a trivial matter. That can be seen by considering the possibility of Kantian contrastivism. Human beings assume that they interact with a world of discrete objects located at points in three-dimensional space and participating in events in a linear time. They assume that phenomena are explicable in terms of a stable set of knowable laws of nature. And they assume that people make decisions for identifiable reasons stemming from their desires. It is central to Kant's philosophy, particularly to the *Critique of Pure Reason*, that we assume these things, and that the assumptions cannot be themselves grounded noncircularly in any more basic evidence or experience. (I'm not doing Kant exegesis; he would have put it differently; that was 230 years ago.) One can argue that physics and psychology give us reasons for hesitating over all of these assumptions. One can do so while also arguing that evolutionary theory supports the idea that something like these assumptions are built into human thinking, and that for most humans in most circumstances thinking without taking these things for granted is not an option.

So consider a person who thinks as people normally do and concludes that the match lit because she struck it. Suppose that she is not familiar with any sophisticated reasons for doubting her natural assumptions. Does she have knowledge? Putting the question contrastively we are asking whether she knows that the match lit because she struck it rather than because it lights at random times, or because she willed it to light. We are accepting that she does not know that the match lit because it was struck rather than because its lighting was part of the computer program that gives an appearance of order to her experience, or because a preordained destiny has laid out the universe in advance, with the striking at time t and the lighting at $t + \varepsilon$. (I think that granting that the person does not have knowledge of these contrasting cases is the right course for a contemporary Kantian. If you disagree, rename the position "contrastive pseudo-Kantianism." The issues remain as hard.) I think it is obvious that if you take two thoughtful intelligent epistemologists at random and ask whether our person has knowledge of why the match struck simply because of her immediate observations and her hardwired Kantian equipment, there will be a fifty percent chance they will disagree. Therefore the issue is not trivial!

Now consider cases like those raised recently in the literature on knowledge and lotteries. You have an appointment with your dentist at 9:00 a.m. tomorrow, and you are an effective planner who is compulsive about appointments. You plan to go to a movie after the appointment. When the dentist's secretary phones you to make sure you have remembered the appointment, you say "sure, I'll be there, barring nuclear war, hurricanes, or heart attacks." You are not saying to her "if there is no nuclear war, etc., I'll be there" but "I'll be there, and I'm assuming that there will be no war or hurricane and that I will not have a heart attack." You are also assuming that you will not have a traffic accident before 9:00 a.m., that your house will not burn down during the night, that no space debris will flatten you, and in fact that none of countless possible preventers will occur. When asked about any of them you will be happy to say that you are assuming they will not happen. And you will be very reluctant, of many of them, to say that you know that they will not occur. For good reason, because you have no evidence that they will not, and many of the factors that would make them occur if they do are random and essentially unknowable. Yet, based on these assumptions, you conclude that you will be in the dentist's waiting room before 9 am, and you do take yourself to know this. (See Hawthorne 2004, especially Chapter 1; the examples in Cohen 2004 have been especially influential.)

I am not going to make any suggestions about the core lottery problem, which is what distinguishes the beliefs that we refrain from calling knowledge even though they follow from things we do know, and why we do know these things even though they are based on unknown assumptions. In a contrastive context these amount to asking why, although you know that you will be at the dentist rather than at the movie at 9:00 a.m., you do not know that you will be at the dentist rather than in an emergency morgue for victims of space debris. That is a hard question, but in asking it we are also seeing how the contrastive point of view takes the bite out of a skeptical paradox. Although it is not clear why we draw the line between knowledge and ignorance where we do, the fact that we fail to have knowledge of some familiar objects relative to some contrasts is quite compatible with our having knowledge relative to other contrasts. You do know that you'll be at the dentist rather than at the movie.

5   FULL CONTRASTS

Ascriptions of non-contrastive knowledge make sense whatever proposition complements the verb. Shakespeare knew that London was in England, he did not know that the 2011 winter Olympics would be in Vancouver, he did not know that $e^{\pi i} = -1$, he did not know that $2 + 2 = 6$. We can even stick in a proposition that cannot be expressed in English, call it p and say that Shakespeare did not know *that*. It is harder to do all this with contrastive knowledge. In particular, it is hard to make sense of contrastive attributions with arbitrary contrasts. Did Shakespeare know that London was in England rather than $3 \times 21 = 64$? Did Shakespeare know that London was in England rather than on Alpha Centauri? The last paragraph of the previous section suggested that skeptical concerns are defanged if we distinguish between knowing that you will be at the dentist rather than at the movie, on the one hand, and knowing that you will be at the dentist rather than dead from the impact of random space debris, on the other. When you say "I know I'll be there" you mean the first. So assume that you do know that you will be at the dentist rather than at the movie, and do not know that you will be at the dentist rather than being at the morgue after a space-debris attack. What follows from this? Is it ruled out that you know that you will be at the dentist rather than at the morgue from dropping in out of curiosity? Is it ruled in, just on general principles rather than as a result of details of your situation, that you know you will be at the dentist rather than at the north pole?

The problem is especially acute if we want to use contrastive knowledge to solve the problem of closure of knowledge under logical consequence. Dretske and Nozick pointed out that one can track p and not track q, even though "if p then q" is a logical truth. They defended the suggestion that the same is true of knowledge, and claimed that this resolves some issues about skepticism. Later philosophers have tended to disagree, although defending closure—the principle that, when one knows propositions $p_1, \ldots, p_n$ and sees how to deduce q from $\{p_1, \ldots, p_n\}$, one comes to know q—has proved to be difficult. (See Hawthorne 2004, especially Chapter 3; also Luper.) My own view is that appealing as the principle is, it is in the end indefensible in full generality, in part because of considerations like those about dentists and morgues. Suppose, however, that one wants to reconstruct closure in terms of contrastive knowledge (see Schaffer 2007). This might be attractive because some putative counterexamples can be defused with well-placed contrasts. Notoriously, you know that you have two hands, and although having two hands entails not being a brain in a vat, you do not know you are not a brain in a vat. But if we qualify the premise to "You know that you have two hands rather than two stumps," the entailment to "You know that you are not a brain in a vat rather than having two stumps" does not seem clearly false.

But it does not seem clearly false because it is so strange that we do not know how to evaluate it. Logical consequence can connect sentences which have so little intuitive connection with one another that they wreak havoc with sensible contrasts. Anyone trying to put together contrastivism and closure will need an attitude to assigning truth values to some very unfamiliar objects.

Even if we decide that clarifying epistemic vocabulary is not alone going to solve problems about closure, we still must face questions about truth values given unrelated contrast propositions. I do not think these questions are insoluble. I propose three principles for handling the issue.

First, the contrast proposition is always false and incompatible with the known proposition. If I know that p rather than q, q is an alternative to p, and because p is true q is false. So we can rule out all of the following: Shakespeare knew that London was in England rather than Stratford was in England, Shakespeare knew that London was in England rather than 2 + 2 = 4, Shakespeare knew that London was in England rather than cats chase mice. The list is easy to extend. In this connection it is worth pointing to an ambiguity. Sometimes, in saying "Shakespeare knew that London was in England rather than Stratford was in England," we might mean something true. That would be when Shakespeare knew that London was in England, did not know that Stratford was in England, and someone had mistakenly asserted that he did know the latter. Then one might correct them by saying "No, it was London that he knew was in England, not (rather than) Stratford." But in saying this one is not ascribing contrastive knowledge in the present sense.

Second, the person has to have discriminated the known proposition from the contrast proposition, by either a perceptual capacity, effective use of evidence, or reliable reasoning. So Shakespeare did not know that London was in England rather than that London within the green belt is in England. Shakespeare did not know that the city from which Elizabeth ruled was in England rather than that the city which Ken Livingston would rule would be in England. Shakespeare did not know that uranium has two isotopes rather than six. He did not know that kings are to be obeyed because of divine command rather than for civic peace. And so on.

Third, we should look for systematic patterns of ascription and denial. Shakespeare knew that London was in England rather than in France, and also that London was in England rather than in Spain, Italy, or Illyria. He did not know that London was in England rather than being in the English sphere of influence or being a British crown dependency. Each of these lists can be continued because the initial contrastive ascription derives from the width of accuracy of Shakespeare's city-identification and city-location capacities, and these capacities make many knowledge contrasts hold and also many fail. When we cannot find such a systematic contrast space we should suspect that the ascription is false.

When these principles do not suggest that someone has contrastive knowledge, then most likely she does not. So the vast majority of random ascriptions of contrastive knowledge are false, just as the vast majority of non-contrastive knowledge claims are. There are many ascriptions that are not settled. But that is as it should be: they have to be settled by data about the particular cases and by an informative epistemology.

6   SUMMING UP: HOW FUNDAMENTAL?

The reason we attribute knowledge is very straightforward. We have reasons to be curious about what aspects of the world creatures have sufficiently accurate information about to guide their actions. Once we have determined this, we can use their actions, including their utterances if they are verbal creatures, as a guide to ours. But what is sufficient to guide one action may not be sufficient to guide another. Tracking-based information is particularly versatile in the variety of actions it can support, but it has its limits: we track some aspects of objects through some possible histories and not through others. As a result, we need a way of relating individuals to facts that does not pretend that the informational stream is wider than it is. So we do two things together. We relate individual agents to the objects they act on in a way that describes the stream of information—the set of action-guiding counterfactuals—that is relevant to actions toward those objects. And we describe the "width" of those streams, the range of similar situations to which an explanation appealing to the same information would apply.

We identify the stream of information by saying what agents know. And we describe its limits by saying what contrasts their epistemic states will support. But the knowledge attribution alone, with no contrast specified, can go some way to describing the limits. This is clearest in cases where there is a close connection with tracking. Then the fact that a knows that p entails that a would believe p under a variety of similar conditions. But only in similar conditions, nearby possible worlds: there is no suggestion that if things had been more than a little different, a would have got a true belief. So the width is vaguely specified by the range of situations S in which the fact that a actually knows that p entails that a has an accurate belief in S. Sometimes an explanation uses an ascription of knowledge in a proposition p to specify a flow of information, and in a rough way its limits, which is then used to explain an action which could be described independently of p: p serves only to pick out the information flow. For example, we can explain why the police managed to arrest a wanted fugitive by the fact that there was a tracking device in a stolen car which he happened to be driving. They knew where the car was, and so they could arrest him. They didn't know where he was.

In this connection it is worth pointing out that explanations by knowledge do not just appeal to knowledge of single propositions. "How could she find you so quickly? Because she knows that forest very well." "Why did Ossie get lost although he knew where the road was? Because he knew that it was north rather than south, but not that it was ten kilometres north rather than five." "Why did Petra succeed in getting the plan approved? Because she knew who to bribe." The crucial phrases here are "knows the forest," "knows where . . .," "knows who . . .". These are all kinds of knowledge that do not center on a particular fact. Instead they centre on a general kind of information possession, a set of counterfactuals, describing the flow of information without singling out one particular origin for it. These less-focused knowledge attributions clearly need specifications of their width. The person who knows a forest very well does not know the location of every mushroom under every tree. So when trying to give the explanation more carefully we say "She knows locations in the forests very well." And in fact we will say "She knows the geography rather than the ecology of the forest very well."

The upshot is that the explanatory work is done by describing systematic information-links between individuals and their environments. We specify these by describing central cases–usually with a "knows" locution—and by describing the limits of the links, usually with contrastive locutions. Sometimes, as I have pointed out, the central factor is not the information on which the action being explained is based. Especially then, we need to determine whether the action on which it is based lies within the limits of the link. Contrastivity is one device for doing this. Often we do not do it contrastively, but leave it to context. A good example of this is "knowing who." Notoriously, one can know who someone is, given the demands of one context, but not given the demands of another. The police may know that the motorist they have pulled over is Jane Doe, because that is what her license truly says, without knowing that she is the notorious graffito artist wanted on three continents. In one sense they know who they have stopped, and in another sense they do not. We can rescue the attribution from context in many ways. One is contrastive: they know that she is Jane Doe rather than Mary Moos (or . . .), but do not know that she is the wanted artist rather than the harmless commuter. They know that they have stopped the transgressor of a minor traffic offense rather than a major traffic offense, but do not know that they have stopped a traffic offender rather than someone wanted by Interpol. We can also use terms that are not explicitly contrastive: they know which citizen they have stopped, but not which criminal. (The awareness of the ambiguity of "knows who" dates to Kaplan 1968; see also Boër and Lycan 1985.)

Contrastive knowledge ascriptions give us information about information links that is essential to using them for explanatory purposes. There are other ways of giving us the information, other explicit linguistic devices and dependence on general contextual inference. There are always other ways of saying things. (Most of the time we can avoid using "knows" if we really want to.) But that does not prevent the information being essential to epistemically based explanations.

REFERENCES

Boër, S., and W.G. Lycan. 1985. *Knowing Who*. Cambridge, MA: MIT Press.
Cohen, S. 2004. Knowledge, Assertion, and Practical Reasoning. *Philosophical Issues* 14(1): 482–491.
Dretske, F. 1981. *Knowledge and the Flow of Information*. Cambridge, MA: MIT Press.
Dretske, F 2000. Epistemic Operators, in *Perception, Knowledge and Belief*. New York: Cambridge University Press: 30–47.
Hawthorne, J. 2004. *Knowledge and Lotteries*. Oxford: Oxford University Press.
Kaplan, D. 1968. Quantifying In. *Synthese* 19: 178–214.
Karjalainen, A., and A. Morton. 2003. Contrastive Knowledge. *Philosophical Explorations* 6(2): 74–89.
Luper, S. The Epistemic Closure Principle. Stanford Encyclopedia of Philosophy. http://plato.stanford.edu/entries/closure-epistemic, accessed 12 Jan 2012
Morton, A. 2009. From Tracking Relations to Propositional Attitudes. *European Journal of Analytic Philosophy* 5(2): 7–18
Morton, A. 2010. Contrastive Knowledge. In *The Routledge Companion to Epistemology*, ed. D. Pritchard and S. Bernecker. New York: Routledge: 513–522.
Nozick, Robert 1981. *Philosophical Explanations*. Oxford, UK: Oxford University Press.

Roush, S. 2005. *Tracking Truth: Knowledge, Evidence, and Science.* Oxford: Oxford University Press.

Schaffer, J. 2005a. Contrastive Causation. *The Philosophical Review* 114(3): 297–328.

Schaffer, J. 2005b. Contrastive Knowledge. In *Oxford Studies in Epistemology 1*, ed. T. Gendler and J. Hawthorne. Oxford: Oxford University Press: 235–273.

Schaffer, J. 2007. Closure, Contrast, and Answer. *Philosophical Studies* 133(2): 233–255.

6 Contrastive Semantics for Deontic Modals

Justin Snedegar

Contrastivism about "ought" is the view that "ought" claims are always relative to a set of alternatives. If "Emmy ought to study" is true, then there is some contextually determined set of alternatives, Q, relative to which it is true. To put the point nonlinguistically: if Emmy ought to study, then there is some set of alternatives out of which Emmy ought to study. Several philosophers have argued that "ought" is contrastive in this way.[1] These philosophers have not, however, extended their contrastive frameworks to other deontic modals, like "must" and "may," and some have explicitly denied that these are contrastive.[2]

In this paper, I motivate and develop a simple contrastive framework for "ought," "must," and "may." In section 1, I motivate contrastivism about "ought" by discussing several puzzles from the literature on deontic modals and deontic logic. This leads to a rough formulation of contrastivism, which will be refined later in the paper. In section 2, I argue that the same sorts of puzzles arise for "must," suggesting that, if contrastivism about "ought" is well motivated, then so is contrastivism about "must." In section 3, I show that, whereas the puzzles do not seem to arise for "may," we can generate similar puzzles for "it is not the case that . . . may . . ." This gives us some evidence that "may" is also contrastive. In section 4, I give a more precise framework for contrastivism about "ought," "must," and "may," and show that all the desired relationships between the modals are preserved. In particular, I show that the framework generates a generalized version of the duality of "must" and "may." I conclude in section 5 by noting some possible implications of adopting contrastivism about deontic modals.

1 MOTIVATING CONTRASTIVISM ABOUT "OUGHT"

In this section I present some puzzles that motivate contrastivism about "ought."[3] The puzzles fall neatly into two groups. First, the Professor Procrastinate puzzle and what I call the Dialogue puzzle have the following form: we have some set of sentences, all of which are intuitively true, but which are apparently contradictory. The task is to reconcile the sentences in the set, so that they can all be true. Second, the Good Samaritan paradox and Ross's Puzzle depend on a particular principle from deontic logic, often called Inheritance. Inheritance says that if p entails q, then if it ought to be that p, then it ought to be that q. The problem is that we can start with intuitively true "ought" sentences, and then use Inheritance to derive intuitively false "ought" sentences. The task is to block this derivation. The solution to both kinds of puzzles is to recognize that "ought" sentences are always relativized to sets of alternatives.[4]

1.1 Reconciling

Consider Professor Procrastinate.[5] She is asked to review a new book in her field because she is by far the most qualified person to write the review. She has time to write, and if she writes it will benefit the field significantly. But she knows that, just because of the way she is, she is extremely likely[6] to (culpably) put off writing if she accepts. If she accepts, the book will sit on her desk, unreviewed for months, while the author's career and the field at large suffer. If she does not accept, someone else, less qualified but more reliable, will be asked to review. The review written by the second choice will be adequate, but not great. But it will be done quickly. Intuitively, then, the following two claims are each true:

(1) Procrastinate ought to accept and write.
(2) Procrastinate ought not accept.

But this is puzzling. How can it be the case that Professor Procrastinate (PP) ought not accept, if she ought to accept and write? The problem is that we have two intuitively true "ought" claims that seem inconsistent.[7]

A naïve view about "ought" says that a sentence containing "ought" has the same semantic content in every context in which it is used or assessed.[8] Clearly, (1) and (2) will cause problems for the naïve view, if we make the plausible assumption that "ought" distributes over conjunction.[9] The goal is to show how (1) and (2) can both be true, but on the naïve view, it seems to follow from (1) that Professor Procrastinate ought to accept, and this seems inconsistent with (2).

I suggest, following Jackson (1985) and Sloman (1970), that a contrastivist view of "ought" can help.[10] The problem, according to the contrastivist, only arises if we ignore the contrast-sensitivity of "ought"; what ought to be is always relative to a set of alternatives. This set is determined by context; generally, the options under consideration in the context will determine the set.[11] If I am trying to decide what to do tonight, and I have been invited to dinner with my sister, to the movies with my wife, and to the bar with my friends, then the set of alternatives will likely be {I go to dinner, I go to the movies, I go to the bar}. If, after some deliberation, I decide that I ought to go to the bar and say so, the contrastivist says that I should be understood to have said something like "I ought to go to the bar out of {I go to dinner, I go to the movies, I go to the bar}."[12]

The contrastivist says that (1) and (2) only seem inconsistent if we ignore the contrast-sensitivity of the "ought" claims. The idea is that (1) and (2) are true relative to different sets of alternatives. (1) is true relative to a set of alternatives like {PP accepts and writes, PP accepts and doesn't write, PP doesn't accept}, whereas (2) is true relative to a set like {PP accepts, PP doesn't accept}. Out of the first set, it is true that PP ought to accept and write, because this is the best option. Out of the second set, the best option is that she doesn't accept, so it ought to be that she doesn't accept. Because the sets of alternatives are different, the claims are not inconsistent. This is easy to see if we make the alternatives explicit, as in: "You ought to accept and write rather than accepting and not writing" or "You ought to decline rather than accept."

What if we assert (1) and (2) in contexts in which the alternatives are the same? That is, what if we assert them in a context in which we should understand (1) as saying "It ought to be that PP accepts and writes out of {PP accepts and writes, PP accepts and doesn't write, PP accepts, PP doesn't accept}," and (2) as saying "It ought to be that PP doesn't accept out of {PP accepts and writes, PP accepts and doesn't write, PP accepts, PP doesn't accept}"? In this case, it seems like (1) and (2) will be inconsistent, even according to the contrastivist. But the goal was to show how (1) and (2) could be consistent.

The first thing to point out is that the contrastivist will likely deny that any context will actually give us sets of alternatives like this. The alternatives should be taken to be mutually exclusive; so we will not get distinct options like "PP accepts and writes" and "PP accepts" in the same set of alternatives. These are not real alternatives, or contrasts, and we will not be deciding between these options. The second thing to point out is that, even if we could get a set of alternatives like this from some context, (2) would be false. Out of this set, the best thing that can happen is for PP to accept and write, so this is what ought to be the case. So it is not a problem that (1) and (2) are inconsistent, if (1) and (2) are taken to be relative to the same set of alternatives.

Contrastivism also offers solutions to several other sorts of puzzles from the literature. Jackson (1985) points out that the contrastivist can easily make sense of apparently contradictory "ought" ascriptions which all seem true. Consider the following dialogue:

Dialogue 1

A: Smith ought to whip his slaves more gently.
B: In fact, he ought to stop whipping his slaves.
C: Really, he ought to free his slaves.
D: He ought to have never owned slaves in the first place.
And so on.

Each of these claims seems true. But they appear to be inconsistent: how can Smith whip his slaves more gently if he doesn't whip them at all, or if he frees them? The contrastivist says that with each ought ascription, the set of alternatives shifts. A's utterance is relative to a set like {Smith whips his slaves more gently, Smith whips his slaves viciously}; B's utterance is relative to a set like {Smith stops whipping his slaves, Smith keeps whipping his slaves}; and so on. Because each ascription is relative to a different set of alternatives, and because this set of alternatives is part of the semantics of the ought ascription, the claims are not inconsistent.

Relatedly, contrastivism easily handles claims that include an explicit "rather than" clause, like the following:

(3) You ought to take the bus rather than driving your SUV, but you ought to bike rather than taking the bus.[13]

Claims like this are perfectly legitimate. But on their face, they can seem inconsistent. The first conjunct says that you ought to take the bus, whereas the second plausibly says that you ought not take the bus. If we assume a naïve semantics for "ought," for example, it is hard to see how a sentence like (3) could be true. More generally, it just isn't obvious how non-contrastive views should handle explicit "rather than" clauses like those in (3). But contrastivism can handle sentences like (3) easily—the "rather than" clause just makes the alternatives explicit.

1.2   Inheritance Puzzles

Next, consider the Good Samaritan puzzle.[14] Suppose you come across a stranger injured on the side of the road, and could easily help him. Then the following claim is true:

(4) It ought to be that you help the injured stranger.

But you help the injured stranger if and only if there is an injured stranger and you help him. Now consider a standard principle from deontic logic:

Inheritance: If p entails q, then if it ought to be that p, it ought to be that q.

So by Inheritance, (5) follows from (4):

(5) It ought to be that there is an injured stranger and you help him.

But presumably this is false. It ought to be that there is not an injured stranger.[15] Contrastivism solves this puzzle by pointing out that the contrast has shifted between (4) and (5).[16] (4) is relativized to a set of alternatives like {you help the injured stranger, you ignore the injured stranger}, whereas (5) is relativized to a set of alternatives like {there is an injured stranger and you help him, there is an injured stranger and you ignore him, there is not an injured stranger}. Out of this latter set, what ought to be the case is that there is not an injured stranger, so (5) is false.

Next, consider Ross's Puzzle, which runs as follows.[17] Suppose you promise your friend that you will mail a letter for him. Then, we can assume

(6) You ought to mail the letter.

But if you mail the letter, then you either mail it or burn it. So by Inheritance, (7) follows from (6):

(7) You ought to either mail the letter or burn it.

But whereas (6) is true, (7) sounds false.[18]

Cariani (2009, forthcoming) proposes an "anti-boxing" semantics for "ought" on which Inheritance fails, thus blocking Ross's Puzzle.[19] On a standard "boxing" theory, what ought to be, in a given world and context, is whatever is the case in the best worlds accessible from that world and context. It is clear, then, why Inheritance is valid on this picture. If it ought to be that p, then p is true in the best worlds. If p entails q, then q will also be true in the best worlds. Thus, it ought to be that q. Cariani's semantics identifies what ought to be the case in a context with the best option in the context, rather than with whatever is true in the best worlds. So if it ought to be that p, then p is the best option. The only way it could be the case that it ought to be that q, which is entailed by p, is if q is tied with p for the best option. In the Ross's Puzzle case, it is not true that "you mail the letter or burn it" is tied for the best option with "you mail the letter"; thus, it is not the case that it ought to be that you either mail the letter or burn it.[20]

The failure of Inheritance on Cariani's semantics leads to a problem. Often, we can make coarse-grained "ought" claims, which are supported by more fine-grained claims that more fully specify what we ought to do. Consider the following inference from (8) to (9), which is perfectly legitimate:

(8) You ought to drive less than 65 mph on this road.
(9) You ought to drive less than 100 mph on this road.

On a boxing view, we can explain this inference using Inheritance. Because "you drive less than 65 mph" entails "you drive less than 100 mph," and because it ought to be that you drive less than 65 mph, it ought to be that you drive less than 100 mph. But Cariani's semantics invalidates Inheritance. In fact, on Cariani's semantics, what ought to be is whatever is the best option. Presumably, if (8) is true, then "you drive less than 100 mph" is not the best option, so (9) comes out false. But intuitively, it is true.

To solve this problem, Cariani (2009) appeals to contrastivism: (8) and (9) are true relative to different sets of alternatives. (8) is true relative to {you drive less than 65 mph, you drive more than 65 mph}, whereas (9) is true relative to {you drive less than 100 mph, you drive more than 100 mph}.[21] Because the more you exceed the speed limit, the worse you are doing, driving less than 100 mph is the best alternative in this set. So if Ross's Puzzle leads us to abandon Inheritance, we can still explain why Inheritance seems to hold in some cases, by appealing to contrastivism. The set of alternatives shifts in a way that seems to validate some Inheritance-like inferences.[22]

Ignoring the contrast-sensitivity of ought ascriptions leads to puzzlement: sets of intuitively true "ought" sentences may appear to be inconsistent, or to entail false "ought" sentences. The contrastivist is able to make the right predictions in these cases. This will serve as the main argument for contrastivism in this paper.

2 "MUST"

In the last section, I motivated contrastivism about "ought" by showing how the theory can solve several puzzling cases from the literature. In this section, I show that these cases also arise for the deontic "must." This suggests that, if contrastivism about "ought" is well motivated, then contrastivism about "must" is, as well. I also discuss an argument from Cariani (2009) that is, at least on a natural reading, a rejection of contrastivism for "must," and show that the same problems which led him to adopt contrastivism about "ought" should lead him to adopt contrastivism about "must."

2.1 Reconciling

First, consider the Professor Procrastinate puzzle, adapted for "must." Suppose the stakes are higher this time around—if the book does not receive a competent review, the author might be denied tenure. Worse, if the book is not reviewed at all, the author will lose her position altogether. Both of the following claims are plausibly true:[23]

(1') Professor Procrastinate must accept and write.
(2') Professor Procrastinate must not accept.

The same problem arises. (1') and (2') both seem true[24], but they are also apparently inconsistent. In the case of "ought," we saw that paying attention to contrasts helped. The same move can help here. (1') is true relative to a set of alternatives like {Procrastinate accepts and writes, Procrastinate accepts and does not write, Procrastinate does not accept}, whereas (2') is true relative to {Procrastinate accepts, Procrastinate does not accept}. Out of the first set, Procrastinate must accept and write. But, given that she will not write, out of the second set, she must not accept. As with "ought" in the last section, I will initially work with a somewhat rough formulation of contrastivism about "must." I offer a more precise framework in section 4.

Next, consider the following dialogue:

> **Dialogue 2:**
> *A:* Smith must whip his slaves more gently.
> *B:* In fact, he must stop whipping his slaves.
> *C:* Really, he must free his slaves.
> *And so on.*

Each of these claims is intuitively true. But again, on a standard semantics for the deontic "must," it seems like they should be inconsistent. If you must p—if p-ing is the only permissible option—then how could it be true that you must q, when q and p are incompatible? Contrastivism about "must" can reconcile the claims. A's claim is true relative to {Smith whips his slaves more gently, Smith keeps whipping his slaves viciously}; B's claim is true relative to {Smith stops whipping his slaves, Smith keeps whipping his slaves}; and so on. Again, paying attention to contrasts can reconcile these intuitively true, but apparently inconsistent, claims. And it can explain why, once B makes her claim, A's claim does not seem true anymore—if A insisted on his claim, he would be mistaken. What has happened is that the conversational context has changed in a way that makes A's utterance false, by taking on a new set of relevant alternatives.[25]

2.2   Inheritance Puzzles

Next, consider the Good Samaritan puzzle, using "must." Suppose you come across an injured stranger on the road, that you could easily help him, that there's no one else around, and so on. Then the following is true (remember that the "must" is deontic, not epistemic):

(4′) It must be that you help the injured stranger.[26]

But you help the injured stranger if and only if there is an injured stranger and you help him. So (5′) follows from (4′):

(5′) It must be that there is an injured stranger and you help him.

But (5′) sounds false. It is not the case that, deontically, there must be an injured stranger. In fact, deontically, it must be that there is no injured stranger (or: it is deontically necessary that there is no injured stranger). Contrastivism about "must" can explain why (4′) is true, while (5′) is false. (4′) is relative to {you help the injured stranger, you ignore the injured stranger}, whereas (5′) is relative to {there is an injured stranger and you help him, there is an injured stranger and you do not help him, there is no injured stranger}. In the former, it is assumed that there is an injured stranger. We make this explicit in (5'), and it introduces the alternative that there just is no injured stranger. Because this state of affairs is better than the state of affairs in which there is an injured stranger and you help him, (5') is false.

Finally, note that a version of Ross's Puzzle also arises for "must." Suppose your friend gives you his rent check to mail; if you don't mail it, then he will be evicted from his apartment. Then the following is true:

(6') You must mail the letter.

But you mail the letter only if you either mail it or burn it. So if we have a principle for "must" which corresponds to Inheritance for "ought," (7') follows from (6'):

(7') You must either mail the letter or burn it.

But (7') sounds false; in fact, I think it is as bad as the corresponding sentence using "ought." Because Ross's Paradox for "ought" motivated Cariani (2009, forthcoming) to reject Inheritance for "ought," he should be motivated to reject the corresponding principle for "must."

By rejecting Inheritance for "ought," Cariani had trouble explaining the legitimacy of inferences from more fine-grained "ought"-claims to more coarse-grained ones. Notice that a similar problem arises, if we reject Inheritance for "must." The inference from "You must drive less than 65 mph on this road" to "You must drive less than 100 mph on this road" is legitimate. Inheritance for "must" gives us an easy way to explain why. But Ross's Puzzle for "must" gave us reason to reject that principle, so we are left looking for an explanation. This is where Cariani (2009) turned to contrastivism for "ought." So this gives us reason to think that "must" is contrastive, as well.

2.3   Rejecting Contrastivism about "Must"

But Cariani (2009) gives an argument that can naturally be taken to show that "must" is not contrastive. He actually focuses on what he calls *prohibition* operators, like "may not" and "prohibited," rather than "must." But I think it's very plausible that "prohibited" just means "must not," which is contrastive only if "must" is contrastive. If you reject this claim, however, you might accept Cariani's argument that prohibition operators are not contrastive, but still be open to contrastivism about "must."[27] Nevertheless, I'll present the argument that Cariani's discussion suggests, even though it's not the one he explicitly gives.

Claims like (10) are perfectly legitimate:

(10) You ought to take the bus rather than driving your SUV, but you ought to bike rather than taking the bus.

This suggests that "ought" is contrastive, because using an explicit "rather than" clause can be understood as a way to make the alternatives explicit. But "rather than" claims with a prohibition operator like "may not" are apparently infelicitous. Cariani (2009) provides the following example:

(11) You may take the bus rather than driving your SUV, but you may not take the bus.

If this is right, it suggests that "must" and "may" are not contrastive, if we think that they are permission and prohibition operators—it would seem that there just are no alternatives there to make explicit.

But I'm not so sure; (12) does not sound obviously contradictory to me, and (13) is obviously fine:

(12) You have to take the bus rather than driving your SUV, but you have to bike rather than taking the bus.[28]
(13) You may take the bus rather than driving your SUV, and you may bike rather than taking the bus.

The sentence Cariani appeals to, (11), does sound contradictory. But notice that it is importantly different than the "ought" claim (10). (10) is analogous to (12) and (13). (14) is an "ought" sentence which is analogous to (11):

(14) You ought to take the bus rather than driving your SUV, but it's not the case that you ought to take the bus.

But this sentence sounds worse than any of (10), (12), or (13), and not clearly better than (11). So this argument that "must" and "may" are not contrastive is not convincing, especially given the apparent parallels between "ought" and "must" with regard to Ross's Puzzle.

3 "MAY"

In the previous two sections, I motivated contrastivism about both "ought" and "must" by showing how the view can solve various puzzles. In this section, I briefly show that some of the same sorts of puzzles—namely, the Professor Procrastinate puzzle and the Dialogue puzzle—do not arise for "may," and offer an explanation for why this is so.[29] I then show that both the Professor Procrastinate and the Dialogue puzzles *do* arise for "it is not the case that . . . may . . ." (from here on, I use "¬may" to abbreviate this), suggesting that "¬may" is contrastive. But of course,

"¬may" is contrastive only if "may" is. This gives us indirect evidence that "may" is contrastive.

3.1   Puzzles for "May"

First, consider the Professor Procrastinate puzzle. Both of the following claims are plausibly true:

> (1″) Procrastinate may accept and write.
> (2″) Procrastinate may [not accept].[30]

The corresponding sentences using "ought" and "must" each seem true, but also appear to be inconsistent. The trouble, if we are trying to generate a puzzle to motivate contrastivism, is that (1″) and (2″) are not inconsistent. We consider performing actions that are neither prohibited nor obligatory all the time. If φ is such an action, then both "s may φ" and "s may [not φ]" will be true. So this version of the Professor Procrastinate puzzle does not provide motivation for adopting contrastivism about "may."

Next, consider the following dialogue.

> **Dialogue 3:**
> *A:* Smith may whip his slaves more gently.
> *B:* Smith may stop whipping his slaves altogether.
> *C:* In fact, Smith may free all of his slaves.
> *And so on.*

All of these claims are true (although they might sound inappropriately weak because of the maxim of quantity). In the case of "ought" and "must," the corresponding claims were also apparently inconsistent. But again, this is not the case with "may"; these claims are all perfectly consistent.

The problem is this: if something ought to be the case, or if something must be the case, then nothing else inconsistent with that thing either ought to be or must be the case in those circumstances. If I ought to/must φ, then it is not true that I ought to/must ψ, where φ and ψ are mutually exclusive (assuming there are no moral dilemmas). So it was easy to get apparently inconsistent "ought" or "must" claims. But it is perfectly consistent for two mutually inconsistent propositions to be such that they may be the case in the same situation (or for two incompatible actions to both be permissible).

3.2   Puzzles for "It Is Not the Case That . . . May . . ."

In the last section, I showed that two of the puzzles that motivate adopting contrastivism about "ought" and "must" do not arise for "may," and explained why. In this section, I show that puzzles do arise for "¬may." And, of course, "¬may" is contrastive if and only if "may" is contrastive.

So these puzzles provide some (indirect) motivation for adopting contrastivism about "may."[31]

Consider, one last time, Professor Procrastinate. The following two claims seem true (using, of course, the deontic "may"):

(1*) It is not the case that Procrastinate may fail to accept and write.
(2*) It is not the case that Procrastinate may accept.

But they are apparently inconsistent; (1*) prohibits Procrastinate from failing to accept and write, whereas (2*) prohibits her from accepting. This is not surprising; (1*) and (2*) just mean the same thing as (1′) and (2′), the "must" analogues. This suggests that "¬may" is contrastive, which, in turn, suggests that "may" is contrastive.

Next, consider the following Dialogue:

Dialogue 4:
A: It is not the case that Smith may fail to whip his slaves more gently.
B: It is not the case that Smith may continue to whip his slaves.
C: It is not the case that Smith may fail to set his slaves free.
And so on.

All of these sentences are plausibly true. But A's claim seems incompatible with B's and C's. Again, contrastivism can help to reconcile these sentences by pointing out that each claim is relativized to a different set of alternatives. Thus, they are perfectly consistent. This provides some motivation for adopting contrastivism about "it is not the case that . . . may . . .," which in turn provides motivation for adopting contrastivism about "may."

I admit that some of the motivations for adopting contrastivism about either "must" or "may" are perhaps less compelling than those for adopting contrastivism about "ought." But I have argued that there is at least some motivation; further, it would be nice to give a unified treatment of deontic modals. If what one ought to do, or what ought to be the case, is contrast-sensitive, it would be a bit surprising if what one (deontically) must or may do, or what must or may be the case, were not similarly contrast-sensitive.

4 CONTRASTIVE SEMANTICS

In the previous three sections, I have worked with a rough, intuitive version of contrastivism about deontic modals. In this section, I propose a more precise contrastive framework that captures all of the right relationships between the modals. I set things up by considering a potential objection: if "must" and "may" are both contrastive, we are in danger of losing the duality of "must" and "may." The framework I propose gives us a contrastivist-friendly understanding of this duality.

4.1   Objection: Lost Duality of "Must" and "May"

"Must" and "may" are duals. Here is the standard formulation of Duality:

Duality: must p $\equiv \neg$[may[$\neg$p]]

This formulation will not do for the contrastivist. Suppose I am trying to decide what to do tonight, and the set of alternatives is {stay in, go to the bar, go to the movies}. And suppose that, as it turns out, I must stay in. So (15) is true:

(15) I must stay in out of {stay in, go to the bar, go to the movies}.

Using Duality, we should be able to replace "must stay in" with "$\neg$[may[$\neg$stay in]]," giving us

(15′) It is not the case that I may [not stay in] out of {stay in, go to the bar, go to the movies}.

The problem is that sets of alternatives are not necessarily closed under negation. "Not stay in" is not in the set of alternatives, and so (15′) does not make sense for the contrastivist, since it cannot be that p out of Q, if p is not in Q. After sketching a contrastive semantics for the deontic modals I've been considering, I'll show how to answer this objection.

4.2   The Semantics

I let the context of utterance of the use of a deontic modal supply (i) a deliberative background, and (ii) a normative background.[32] The deliberative background provides a set of alternatives Q, to which the "ought," "must," or "may" claim is relative. The normative background provides (a) a ranking < of the alternatives, and (b) a selection function, L, which selects the lowest-ranked permissible alternative from Q, or the "least you can do."[33] I leave it open both how the ranking is determined and how the least you can do is determined. Suppose we have a ranking of the alternatives in Q, and that L selects, as the least you can do, alternative a. Then we can define "ought," "must," and "may" as follows:[34]

May: may p out of Q $=_{df}$ p is ranked at least as highly as a
Must: must p out of Q $=_{df}$ p = a, and there is no q $\neq$ p in Q which is ranked as highly as p

Ought: ought p out of Q $=_{df}$ there is no q ≠ p in Q which is ranked as highly as p[35]

Intuitively, an alternative may be the case when it is the least you can do or better; an alternative must be the case when it is the only alternative that may be the case; and an alternative ought to be the case when it is the best alternative.[36]

Suppose context provides us with a set of alternatives, Q={a, b, c, d}, ranked in that order, and that L selects c as the least you can do. Then by our semantics, it may be that a, it may be that b, and it may be that c, out of Q. And because a is ranked the highest, it ought to be that a. So we can truly say, "It ought to be that a (out of Q), but it may be that b or c instead." This seems right. We often say things along these lines. Suppose Emmy has gotten a bonus at work. Then the following is perfectly appropriate, I think: "You ought to save that money, but you may take your husband to a nice dinner, instead." We're saying that saving the money would be the best thing to do, but it would be permissible to spend the money on her husband.[37] But because there is more than one permissible option (more than one alternative is ranked at least as highly as the alternative selected by L), nothing must be the case. This is intuitive. If there is more than one permissible option, then there is nothing which must be the case; note that the following sounds inappropriate: "It may be that a or b (out of Q), but it must be that a."[38]

This picture assumes that if anything may be, must be, or ought to be the case out of a set of alternatives, L will select something as the least you can do; that is, it assumes that there is always some permissible option. If we are considering a set of alternatives for which no option is permissible, then L will not select anything. It follows that nothing will be such that it may be the case, must be the case, or ought to be the case, out of that set of alternatives. It isn't obvious that there will be contexts that provide such a deliberative background, but if there are, these results seem correct.[39]

4.3   Relative Strengths

An adequate semantics for deontic modals should preserve the following relative strengths: "must" is stronger than "ought," and both are stronger than "may." Using this framework, we get all the right results: it is easy to see that "must p out of Q" entails "ought p out of Q," but that the converse does not hold; further, both entail, but are not entailed by, "may p out of Q."

4.4   Recovering Duality

In the framework I have sketched above, "must" and "may" stand in the following relationship: "must p" means that L selects p, and < does not rank any alternative higher than p. That is, p is the only permissible alternative. So if "must p" is true, then "may q" is false, for any q not identical with p. This suggests the following generalized version of Duality:

> **Contrastivist Duality (CD):** must p out of Q ≡ for any q in Q not identical with p, ¬[may q] out of Q.

CD may initially appear very dissimilar to standard Duality. But notice that standard Duality actually *falls out* of CD as a special case when the only options in Q are p and its negation, ¬p. For example, suppose the set of alternatives Q is {go out, do not go out}. And suppose that I must go out, out of Q. Because "not go out" is an element of Q not identical with "go out," by CD, it follows that it is not the case that I may [not go out] out of Q; that is, "must p out of Q" entails "¬[may[¬p]] out of Q." Now suppose that it is not the case that I may [not go out], out of Q. Then because "not go out" is the only alternative in Q, not identical with "go out," it follows from CD that I must go out. That is, "¬[may[¬p]] out of Q" entails "must p out of Q."

5 CONCLUSION

I have argued that we should take seriously the idea that "ought," "must," and "may" are contrastive. Contrastivism allows us to solve various puzzles surrounding these modals in a nice, unified way. Further, I have shown that it is possible to give a simple, unifying semantic framework that captures all of the desired relationships between the modals.

Adopting contrastivism about deontic modals could have interesting implications elsewhere. For example, I have shown that we would need to replace the standard understanding of the duality between "must" and "may" with a generalized version. It is plausible that other principles and axioms from deontic logic would have to be generalized, or otherwise altered. It could also have implications elsewhere in normative philosophy, because most ethical theories are concerned with what we ought to do, or what we must do, or what we may do. If it turns out that our use of these deontic modals is contrastive, it would not be surprising if this reflected important features of the role of alternatives in our moral thinking. Finally, many philosophers have argued that there is some tight connection between "ought" and reasons. Although adopting contrastivism about deontic modals may not *force* one to adopt a contrastive picture of reasons, it does suggest that it would be interesting to investigate the prospects for such a view.[40] I do not explore any of these implications here, although I think they might prove to be interesting and fruitful areas for future research.[41]

NOTES

1. See Sloman (1970), Jackson (1985), Jackson and Pargetter (1986), Finlay (2009), Cariani (2009), and Cariani (forthcoming).
2. I have in mind Finlay (2009), who offers a non-contrastive semantics for "must" and "may," and Cariani (2009) who gives an argument which can naturally be taken to be to the conclusion that modals are not contrastive (although he tells me in conversation that he doesn't want to commit himself to that conclusion).
3. I do not claim that the *only way* to solve these puzzles is to adopt contrastivism—only that contrastivism provides a nice, unified solution.
4. In fact, there are two distinct features that a contrastivist might accept, and these two features let the contrastivist solve different puzzles, and solve the same puzzles in different ways. The first feature is that the sets of alternatives to which an "ought" sentence is relativized need not be *exhaustive* of all of logical space—in this way, a contrastivist view is similar to Lewis's (1996) relevant alternatives theory of knowledge. The second feature is that a set of alternatives might divide up the alternatives in different ways. For example, one set might be {I go to the bar, I stay home}, while a different set might be {I go to the bar, I stay home and watch a movie, I stay home and work on my dissertation}. Cariani (forthcoming), following Yalcin (2011), calls this feature *resolution-sensitivity*: we divide up the space of possibilities at different resolutions, or at different levels of fine-grainedness. This feature comes with relativizing "ought" sentences to *questions*. See Schaffer (2007) and Groenendijk and Stokhof (1997) for relevant discussion.
5. Jackson (1985) discusses this case, and the contrastivist solution. See also Jackson and Pargetter (1986) and Cariani (forthcoming).
6. In Jackson and Pargetter (1986), we are told that Procrastinate *will not* write, rather than that she is *extremely unlikely* to write. Some people have the intuition that, if we set things up in Jackson and Pargetter's way, (1) isn't clearly true because accepting and writing seems psychologically impossible for Procrastinate. This is why I leave it open that Procrastinate *might* write (although it is extremely unlikely).
7. *Possibilists* deny that (2) is true. See Jackson and Pargetter (1986) for arguments against this view.
8. Here, I ignore issues surrounding the so-called "ought to be" versus "ought to do" distinction. The view that there are two senses of "ought," one which corresponds to the "ought to be" and one which corresponds to the "ought to do," but each of which has a constant semantic content in every context of use and assessment, will qualify as a naïve view. For recent discussion of this issue, see, e.g., Schroeder (2011), Ross (2010), and Finlay and Snedegar (manuscript).
9. The contrastivist semantics I offer rejects this assumption, but in a principled way. See also Jackson (1985).
10. A non-contrastivist contextualist or relativist view could not solve this puzzle, at least not in any way that is obvious to me. See Kolodny and MacFarlane (2010), Finlay and Björnsson (2010), and Dowell (manuscript).
11. See Sloman (1970) for more discussion of how different sets of alternatives might be determined.
12. This formulation follows Sloman (1970) and Jackson (1985) closely.
13. See Cariani (2009). And compare Schaffer (2008) on contrastivism about knowledge.
14. This puzzle is presented in Prior (1958).
15. Some have offered a scope solution to this kind of puzzle; see Sinnott-Armstrong (1985) for discussion. Because of puzzles like this one, and Chisholm's Paradox

(Chisholm (1963)), it is standard in deontic logic to appeal to a conditional "ought," "it ought to be that φ given ψ." This allows us to account for contrary-to-duty paradoxes, like Chisholm's paradox, and other related puzzles. It would be interesting to see if we could handle all of the puzzles presented here using a conditional "ought," but I don't do that here. I only note that appealing to a conditional "ought" does not seem to help with the Professor Procrastinate puzzle, because this puzzle does not appear to depend on any kind of conditional. Contrastivism, then, offers a unified solution to the various puzzles, while the conditional "ought" does not obviously do so.

16. See Jackson (1985).
17. The puzzle is presented in Ross (1941).
18. Most people offer a pragmatic explanation of Ross's Puzzle, along Gricean lines (Grice (1989)). See Føllesdal and Hilpinen (1971) and Wedgwood (2006). But Cariani (2009, forthcoming) argues that this explanation does not work, because the relevant implicatures in Ross's Puzzle behave differently than the sorts of implicatures under which these authors want to subsume them.
19. Cariani calls his semantics "anti-boxing" because he does not treat "ought" as a quantifier over worlds, or "box," as the standard semantics does.
20. The options in a set of alternatives are also generally taken to be mutually exclusive. If we accept that, then of course we won't even have a set of alternatives that includes both "you mail the letter" and "you mail the letter or burn it".
21. Cariani (2009) also attempts to show how the truth of (8) *explains* the truth of (9), but this is somewhat tangential to my purposes. What I've said so far should make it clear why Cariani adopts contrastivism—doing so is necessary for even saying that (9) is *true*.
22. For a more detailed version of this argument, see Cariani (2009).
23. The intuitions here might not be as robust as in the "ought" case, but I can hear (1) and (2) as both true. Throughout this section, feel free to substitute "has to" or "is required to" for "must." These are equivalent, for my purposes.
24. Steve Finlay suggests in conversation that it would be strange to assert both, but points out that it seems appropriate to assert *either*, as long as we do not go on to assert the other. I think this is enough to support the claim that both seem true, although it does point to a disanalogy between the "ought" case and the "must" case, because in the "ought" case, it seems fine to assert both at the same time.
25. Compare Jackson (1985).
26. The epistemic reading of "must" is more natural than the deontic reading, which I intend. Perhaps it would be better to replace "it must be that" with "it is deontically necessary that," assuming that "must" is a deontic necessity operator. Thanks to Steve Finlay here.
27. In fact, this is precisely what Cariani wants to do (personal communication); he wants to remain open to the possibility that "must" is contrastive.
28. I am assuming that "have to" is synonymous with "must" here. "Have to" just sounds much more natural to me than "must."
29. The puzzles which involve Inheritance do seem to arise, although it is hard to evaluate whether or not these really support contrastivism about "may," so I leave them out.
30. I'll use brackets to distinguish "may [not φ]" from "[may not] φ," because "may not" is usually synonymous with "must not" in English. What I have in mind here is not prohibition, but rather the permissibility of refraining.
31. The sentences in this section come out a bit stilted, because "it is not the case that" isn't really a common phrase in everyday English. Cariani (personal communication) suggests that the same kinds of arguments could be run using "I doubt that . . . may." I think that's an interesting and plausible suggestion, but I don't have the space to explore it here.

32. This framework is similar to the semantics developed by Cariani (2009), although his normative base provides a selection function which selects the most highly ranked alternative, instead of the "least you can do." But in his more recent work, Cariani (forthcoming), he (independently of my work on the issue) develops a semantic theory which *does* make use of this sort of "least you can do" bar, although he puts it to different use than I do. Both his theory and my theory have roots in Kratzer's (1981) semantics for modals, although we both modify the semantics beyond anything Kratzer would likely accept.
33. See McNamara (1996c), McNamara (1996b), and especially McNamara (1996a) for discussion of the "least you can do," and how to incorporate it into a formal semantics.
34. This picture needs to be amended to allow for information-relativity regarding the modals, but I don't have the space to take up that issue here. See Kolodny and MacFarlane (2010), Finlay and Björnsson (2010), and Dowell (manuscript).
35. I am assuming here that there are no conflicts about what ought to be the case or what must be the case. I think the semantics could be amended to allow for such conflicts—the first move, at least, would be to change the definition of "ought" and the second half of the definition of "must," to say that φ is ranked *at least as highly* as every other alternative.
36. Here I am assuming that < ranks the alternatives from best to worst, and leaving it open what constitutes the best. This may be different in different contexts.
37. McNamara (1996c) argues convincingly that it isn't always true that we must do what we ought to do.
38. One interesting complication, which I don't have the space to take up here, is that if a, b, and c are the only permissible options, we might want to say "It must be that (a ∨ b ∨ c)."
39. It is at least arguable that in a situation in which all the alternatives are bad, the least bad alternative still (say) ought to be the case. It might sound strange to say that, in such a situation, some bad alternative, even if it is least bad, is selected by L as the lowest-ranked *permissible* alternative. But I do not mean to use "permissible" in any morally loaded way. In some contexts, the lowest-ranked permissible alternative, in my sense, might turn out to be impermissible, according to some set of moral standards.
40. Sinnott-Armstrong (2006, 2008) argues that reasons for belief are contrastive, and suggests that *all* reasons are contrastive. I take up contrastivism about reasons in Snedegar (2012).
41. Thanks to audiences at USC, Brandeis, UC-Berkeley, and the Rocky Mountain Ethics Congress at the University of Colorado. Thanks also to Ben Lennertz, Martijn Blaauw, Stephen Finlay, and Fabrizio Cariani for comments and discussion. Thanks to Walter Sinnott-Armstrong for discussion and for recommending this paper for inclusion in this volume. And finally, my greatest debt is to Mark Schroeder for many hours of discussion and many rounds of comments.

REFERENCES

Cariani, F. 2009. The Semantics of "Ought" and the Unity of Modal Discourse. PhD diss., University of California–Berkeley.
Cariani, F. Forthcoming. "Ought" and Resolution Semantics. *Noûs*.
Chisholm, R. 1963. Contrary-to-Duty Imperatives and Deontic Logic. *Analysis* 24: 33–36.
Dowell, J. Manuscript. A Flexible Contextualist about "Ought."

Finlay, S. 2009. Oughts and Ends. *Philosophical Studies* 143: 315–340.

Finlay, S., and G. Björnsson. 2010. Metaethical Contextualism Defended. *Ethics* 121(1): 7–36.

Finlay, S., and J. Snedegar. Manuscript. One "Ought" Too Many.

Føllesdal, D., and R. Hilpinen. 1971. Deontic Logic: An Introduction. In *Deontic Logic: Introductory and Systematic Readings*, ed. D. Føllesdal and R. Hilpinen, 1–35. Dordrecht: D. Reidel Publishing Company.

Grice, H. 1989. Logic and Conversation. In *Studies in the Way of Words*, 22–40. Cambridge, MA: Harvard University Press.

Groenendijk, J., and M. Stokhof. 1997. Questions. In *Handbook of Logic and Language*, ed. J. van Benthem and A. ter Meulen, 1055–1124. Amsterdam, The Netherlands: Elsevier Science Publishers.

Jackson, F. 1985. On the Semantics and Logic of Obligation. *Mind* 94(374): 177–195.

Jackson, F., and R. Pargetter. 1986. Oughts, Options, and Actualism. *Philosophical Review* 95(2): 233–255.

Kolodny, N., and J. MacFarlane. 2010. Ifs and Oughts. *Journal of Philosophy* 107(3): 115–143.

Kratzer, A. 1981. The Notional Category of Modality. In *Words, Worlds, and Contexts*, ed. H.J. Eikmeyer and H. Rieser, 38–74. Berlin: de Gruyter.

Lewis, D. 1996. Elusive Knowledge. *Australasian Journal of Philosophy* 74: 549–567.

McNamara, P. 1996a. Doing Well Enough: Toward a Logic for Common-Sense Morality. *Studia Logica* 57(1): 167–192.

McNamara, P. 1996b. Making Room for Going beyond the Call. *Mind* 105(419): 415–450.

McNamara, P. 1996c. Must I Do What I Ought (or Will the Least I Can Do Do)? In *Deontic Logic, Agency and Normative Systems*, ed. M.A. Brown and J. Carmo, 154–173. Berlin: Springer-Verlag.

Prior, A. 1958. Escapism: The Logical Basis of Ethics. In *Essays in Moral Philosophy*, ed. A. Melden, 135–146. Seattle, WA: University of Washington Press.

Ross, A. 1941. Imperatives and Logic. *Theoria* 7: 53–71.

Ross, J. 2010. The Irreducibility of Personal Obligation. *Journal of Philosophical Logic* 39(3): 307–323.

Schaffer, J. 2007. Knowing the Answer. *Philosophy and Phenomenological Research* 75(2): 383–403.

Schaffer, J. 2008. The Contrast Sensitivity of Knowledge Ascriptions. *Social Epistemology* 22(3): 235–245.

Schroeder, M. 2011. "Ought," Agents, and Actions. *Philosophical Review* 120(1): 1–41.

Sinnott-Armstrong, W. 1985. A Solution to Forrester's Paradox of Gentle Murder. *Journal of Philosophy* 82(3):162–168.

Sinnott-Armstrong, W. 2006. *Moral Skepticisms*. Oxford: Oxford University Press.

Sinnott-Armstrong, W. 2008. A Contrastivist Manifesto. *Social Epistemology* 22(3): 257–270.

Sloman, A. 1970. "Ought" and "Better." *Mind* 79(315): 385–394.

Snedegar, J. 2012. *Contrastive Reasons*. PhD diss., University of Southern California.

Wedgwood, R. 2006. The Meaning of "Ought." *Oxford Studies in Metaethics*, vol. 1, ed. R. Shafer-Landau, 127–160. Oxford: Oxford University Press.

Yalcin, S. 2011. Nonfactualism about Epistemic Modality. In *Epistemic Modality*, ed. A. Egan and B. Weatherson. Oxford: Oxford University Press.

7 Free Contrastivism

Walter Sinnott-Armstrong

1 CONTRASTIVISM

Many central issues in philosophy concern reasons. Epistemology is about reasons to believe (or disbelieve) certain propositions. Ethics is about reasons to do (or not do) certain actions. Metaphysics and philosophy of science often focus on causation and explanation, which involve reasons why certain events do (or don't) happen.

All of these investigations can benefit from contrastivism. A contrastivist view of a concept holds that all or some claims using that concept are best understood with an extra logical place for a contrast class. As a universal theory of reasons, contrastivism about reasons claims that a reason for something is always a reason for that thing as opposed to some contrast. The point is not that there is a reason for a contrastive proposition ("one thing rather than another") but, instead, that the reason favors one thing and disfavors others. It is the reason, not the proposition, that introduces the contrast.

This abstract view applies to various kinds of reasons, including epistemic reasons to believe. In the classic case from Dretske (1970), a father takes his daughter to the zoo, and, when she asks him what kind of animal is in a certain cage, he answers, "That's a zebra." Does he know (or is he justified in believing that) it is a zebra? An epistemological contrastivist would respond that the father knows that it is a zebra in contrast with an aardvark, bear, camel, duck, elephant, and so on for other normal animals. Still, the father does not know that it is a zebra as opposed to a mule painted to look just like a zebra, a robot with zebra fur on top of metallic parts, a perfect holographic image of a zebra, or an image in "The Matrix." Why not? Because the father's visual experience enables him to distinguish zebras from aardvarks and all other members of the first contrast class (so he can rule out those other animals), but he still cannot distinguish zebras from (or rule out) members of the second contrast class. The need for such contrasts in epistemology is supported forcefully by Dretske (1970, 1972), Schaffer (2004a, 2004b, 2005a, 2008), and Karjalainen and Morton (2003, 2008). Sinnott-Armstrong (2006) extends this approach into moral epistemology.

Other philosophers apply contrastivism to explanation, which involves reasons why things happen. Humidity explains why it rains instead of not precipitating at all, but temperature explains why it rains instead of snows. Van Fraassen (1980) and Lipton (1991) show how to build such contrasts into an illuminating and fruitful theory of explanation in general. This approach is extended from causal explanations to causation by Schaffer (2005b).

The same point applies to reasons for action. My reason to cook a cake on my son's birthday as opposed to the day after his birthday is that his party is on his birthday, whereas my reason to cook a cake instead of a pie on his birthday is that cake is traditional on birthdays, and he likes cake. There are different reasons for different contrasts, as before. Nobody has yet developed a contrastivist theory of practical reasons to act, but it seems natural (see Sinnott-Armstrong 2006, 112–117).

Why adopt such a convoluted and complicated view? Contrastivism is justified by its ability to illuminate examples, as in the cases cited, and to resolve or avoid puzzles and paradoxes.

One prominent puzzle in epistemology is the challenge of skepticism. Do I know that I have hands? Philosophers have argued for a long time about how to answer this simple question, because it is hard to explain how I can know that I have hands when I cannot rule out incompatible skeptical hypotheses, such as that I am a disembodied brain stimulated to see images of hands. Contrastivists can say that I know that I have hands in contrast with claws or wings, but I do not know that I have hands in contrast with images of hands created by stimulation of my disembodied brain in a vat. If contrastivists also deny that either contrast is the one that determines whether or not I just plain know that I have hands, then they can refuse to answer the simple traditional question of whether I just plain know that I have hands. In that way, they can resolve the skeptical paradox and avoid ancient disputes (see Sinnott-Armstrong 2004, 2008).

Contrastivism can also help to resolve other puzzles in the philosophy of science, such as Goodman's grue paradox (Goodman 1955). Do our past visual experiences of emeralds give us evidence that emeralds are green even though those experiences are also compatible with the contrary hypothesis that emeralds are grue? This question is puzzling because it asks simply whether past experiences are evidence that emeralds are green. The puzzle dissolves (or is reduced) when we add contrast classes: past experiences give us evidence that emeralds are green as opposed to blue, but those experiences do not give us any evidence that emeralds are green as opposed to grue.

A metaphysical puzzle where contrastivism helps is mental causation. Mental states or events seem to have physical effects, such as when our intentions or choices seem to cause our bodies to move. How can such "downward causation" work? The answer lies in multiple realizability. Suppose that a mental state (M_1) could be realized in any of several different brain states (B_1, B_2, B_3, and so on), but it happens to be realized in B_1 on a particular occasion. Now suppose that a physical state (P_1) results and would have resulted even if M_1 had been realized in any of the contrasting brain states (B_2, B_3, and so on) instead of B_1, but P_1 would not have resulted if a contrasting mental state M_2 had occurred instead of M_1. In this case, the occurrence of the mental state M_1 as opposed to mental state M_2 is what causes P_1 rather than contrasting physical state P_2. It would be inaccurate to respond that the occurrence of the brain state B_1 as opposed to B_2 causes P_1 rather than P_2, because P_1 would occur even with B_2 instead of B_1, so the requisite counterfactuals do not hold for "B_1 as opposed to B_2 causes P_1 as opposed to P_2" as they do for "M_1 as opposed to M_2 causes P_1 as opposed to P_2." The point is that the mental kind description rather than the neural kind description can capture the relevant level of generality for causal laws as well as explanations. For example, if fear (as opposed to joy) causes increased (as opposed to baseline) blood flow in the amygdala as well as movement away (rather than toward) a snake, regardless of which brain state among several possibilities happens to realize that fear, then fear rather than the particular brain state that realizes fear on an occasion is what causes the movement away from the snake on that occasion (see also Craver 2007, 202–211, 223–224). In something like this way, contrastivist accounts of causation can illuminate apparent causal relations from mind to body and might also help to defend a qualified version of the commonsense view that our choices affect what we do. And this account works even if all mental events are completely constituted by physical events on every particular occasion.

Next, consider moral dilemmas. When Sophie is taken to a Nazi concentration camp with her two children, the guard tells her that she must choose one child to die and one to live in the camp, and both will be killed if she refuses to choose. Does she have a reason to choose her daughter? This question is puzzling without contrasts, but a contrastivist can say that Sophie has a reason to choose her daughter instead of neither child even if she has no reason at all to choose her daughter instead of her son. This account does not make her choice any easier, of course, but it avoids contradiction and clarifies what she does and does not have reason to do. Thus, contrastivism would have helped me in Sinnott-Armstrong (1988).

In all of these cases, puzzles arise when philosophers pose questions (or make claims) about reasons without specifying any contrast class. They argue about the reason for X without specifying what contrasts with X. The puzzles can be avoided or solved by insisting on filling out the contrast classes at least when the need arises.

Traditionalists often respond that these philosophical issues survive for unqualified claims about reasons independently of any contrast class. Such claims about reasons can be understood as presupposing that a certain contrast class is the relevant one for determining whether someone really and truly has a reason to believe or do something. The assumption that one contrast class is the relevant one can then be seen as the source of the trouble. The trouble can be avoided by rejecting that assumption along with questions about who really and truly has a reason without qualification.

Instead of taking sides, contrastivists can then suspend belief about which unqualified reason claims are true. This kind of contrastivism can be called Pyrrhonian because Pyrrhonian skeptics suspend belief about unresolvable philosophical quarrels. This position is admittedly more theory-laden than most Pyrrhonians would like, but it shares their doubts about many of the questions that baffle traditional philosophers. The benefits of this version of contrastivism come not from picking sides but from clarifying issues and showing how to avoid ancient disputes. Contrastivists of this sort dissolve rather than solve traditional philosophical issues.

We should expect this general pattern to recur in accounts of freedom. After all, whether one acts or wills freely depends on the reasons that explain or cause one's act or will. These reasons, explanations, and causes depend on contrast classes. As a result, freedom also depends on contrast classes. Indeed, freedom is contrastive in more ways than one. When someone asks whether an agent is free, we need to ask at least two questions about contrasts: Free from what? Free to do what?

2   FREE FROM WHAT? CONTRASTING CONSTRAINTS

Most traditional views of freedom in philosophy are reactions to the classic problem of determinism. This argument poses that problem simply:

1. Every act is (fully) determined by preceding causes.
2. If any act is (fully) determined, then its agent is not free (at all).
3. If any agent is not free (at all), then that agent is not responsible (at all).
4. If any agent is not responsible (at all), then that agent should not be punished (at all).
5. If any agent is punished who should not be punished (at all), then the punisher owes that agent an apology and compensation.
6. Therefore, we owe an apology and compensation to every rapist and murderer whom we ever punished.

Almost nobody wants to accept this conclusion, of course. The problem is that the argument is valid, so the conclusion cannot consistently be avoided without denying a premise, and it is not clear which premise to deny.

Libertarians who allow contra-causal freedom deny premise 1. Compatibilists about freedom deny premise 2. Compatibilists about responsibility but not about freedom deny premise 3. Hard determinists and hard incompatibilists deny premise 4 or premise 5. I have always felt torn between compatibilism and hard determinism. Contrastivism provides a way to have it both ways at once.

The key to the contrastivist solution is to insert a place for a variable into the account of freedom. Specifically, freedom is always freedom *from* a certain range of constraints. When someone asks whether an act or person is free, instead of answering the question directly, we often need to ask, "Free from what?"

This simple thesis is not new. Many philosophers throughout history have made roughly the same point in different ways. Here are some of my favorites:

> David Hume: "Few are capable of distinguishing betwixt the liberty of *spontaneity*, as it is call'd in the schools, and the liberty of *indifference*; betwixt that which is *opposed to* violence and that which means a *negation of* necessity and causes." (1739–1740, 407; last two emphases added to indicate contrasts)

> J.L. Austin: "While it has been the tradition to present this ['freedom'] as the positive term requiring elucidation, there is little doubt that to say we acted 'freely' (in the philosopher's use, which is only faintly related to the everyday use) is to say only that we acted *not* un-freely, in one or another of the many heterogeneous ways of so acting (under duress or what not). Like 'real', 'free' is only used to rule out the suggestion of some or all of its recognized antitheses." (1961, 180)

> Joel Feinberg: "It is useful to interpret these singular judgments [that he is free] in terms of a single analytic pattern with three blanks in it: _____ is free from _____ to do (or omit, to be, or have) _____." (1973, 3–4)

> Peter Unger: "Under what conditions is a person *free to do* a certain thing? He must be *free from* a plenitude of factors, say, 'constraining,' 'binding,' 'preventing' factors." (1984, 57)

Thus, contrastivism about freedom is at least not too idiosyncratic.

Some critics object that only negative freedom is freedom from constraints, whereas positive freedom is freedom to do something rather than freedom from anything. This distinction, however, is best understood as a distinction between different kinds of constraints (as shown by Feinberg 1973, 5–7). A pauper is not free from legal constraints against stealing a yacht but is free from legal constraints against buying a yacht. Nonetheless, he is still not free from financial constraints against buying a yacht, because he has no money. This pauper, then, is said to have negative but not positive freedom to buy a yacht. But why does he lack positive freedom? Because the pauper is constrained by law against a conjunctive act: taking the yacht without paying for it. Even apart from law, the seller would stop the pauper from taking the yacht without paying. Thus, the pauper is constrained by his lack of money. Now suppose that this constraint disappears, because the pauper wins a lottery, but he is still terrified of water, so he cannot bring himself to buy a yacht. Then he is constrained by hydrophobia, but not by his bank account or the law. He has negative freedom from some constraints (both legal and financial), but he still lacks positive freedom because of other constraints (psychological). There is, thus, a distinction between positive freedom and negative freedom, but both kinds of freedom are freedom from some range of constraints. The distinction lies in the different kinds of constraints, and all such constraints can replace the variable in contrastivist accounts of freedom from.

Why accept this contrastivism? Part of the purpose of a theory of freedom is to help us understand common language and common concepts. Accordingly, one argument for contrastivism about freedom appeals to common language:

> "Those mints are free. Take one if you want."
> "I am free to see you now. Come right in."
> "That table by the window is free. Let me seat you there."
> "You are free to join us if you want."
> "America is a free country."

The contrastivist account can easily explain why we naturally use the word "free" in these expressions. To say that the mints are free is to say that they are free from financial cost, so you are not constrained from taking one even if you do not have any money. To say that I am free to see you now is to say that I am free from conflicting obligations and not constrained by previous commitments. To say that the table by the window is free is, similarly, to say that no reservation constrains you from sitting there or the waiter from seating you there. To say that you are free to join us is to say that your joining us is free from any constraint of etiquette, such as the rule that you should not join a group when the group does not want you to join it. To call America a free country is to say that its citizens are not constrained by law or government in many of the ways that people in other countries are constrained.

Non-contrastive accounts of freedom have no good way to deal with such common expressions. What do free mints, free tables, free countries, and so on have to do with whether anyone's actions are or are not determined by a prior cause (as on libertarian accounts of freedom) or with whether or not anyone is reasons-responsive (as on some compatibilist accounts of freedom)? Such non-contrastive accounts must dismiss this common language as deviant or multiply ambiguous. These dodges introduce complexities in linguistic theory, so they should be avoided for the sake of simplicity if possible. Contrastive accounts show how to avoid such unconstrained ambiguity. Hence, a contrastive account has definite advantages from a linguistic point of view.

Opponents might respond that free mints, tables, and countries are beside the point, because our main concern here is freedom of action and will. But contrastive accounts of freedom also have advantages for understanding freedom of action and will. In particular, they resolve or avoid the problem of determinism. To see how, consider these contrast classes that specify what someone is free from:

> Free from causation = not determined by any cause of any kind.
> Free from (external) constraint = not prevented by (external) physical barriers or forces (such as ropes, bars, weights, pushing, . . .).
> Free from (internal) compulsion = not due to volitional or emotional mental illness (such as addiction, obsession, compulsion, phobia, . . .).
> Free from ignorance = not due to delusion or reasonable mistake.
> Free from coercion = not required in order to avoid excessive costs or risks (such as from personal threats or impersonal circumstances).
> Freedom from prohibition = not forbidden by law or morality.
> Free from excuse = free from constraint, compulsion, ignorance, coercion, and prohibition.

Some of these conditions (such as coercion and prohibition) might be called justifications rather than excuses, and the differences among them matter in some ways. Still, for simplicity, I will here lump them together under an extended notion of excuse.

On the contrastivist account, the listed notions of freedom are all legitimate and distinct. An agent who is not free from causation still might be free from all constraint, compulsion, ignorance, coercion, and prohibition of excusing kinds. Conversely, an agent who is not free from coercion or ignorance still might be free from causation. Similarly, an agent who is free from coercion might not be free from internal compulsion or ignorance, and vice versa.

All of these distinctions would be just so much rigmarole if they did not help somehow, but they do. For one thing, the distinction between being free from causation and being free from excuse helps resolve or avoid the problem of determinism that opened this section. Premise 2 in that argument claims that any determined act is not free. That premise is unquestionable if it refers to freedom from causation as defined above, because then all it claims is that any act that is determined by causation is not free from causation. Nonetheless, premise 2 is eminently questionable if it refers to freedom from excuse, because then it claims that any determined act is not free from excuse. That claim begs the question against compatibilists. Because they deny that premise, it needs to be supported by some independent argument to show why all forms of causation should count as excuses that remove responsibility. Nothing like that has been shown yet in the argument above.

So let's suppose that premise 2 is interpreted in terms of freedom from causation. What about the next premise? Premise 3 in the above argument claims that any agent who is not free is also not responsible. That premise is unquestionable if it refers to freedom from excuse, because the relevant kinds of constraint, compulsion, ignorance, coercion, and prohibition are defined to include only those kinds that excuse in the broad sense that removes responsibility. Hence, anyone who is not free from excuse has an excuse and, hence, is not responsible. But if premise 2 refers to freedom from causation and premise 3 refers to freedom from excuse, then the argument equivocates and is invalid.

To avoid equivocation, premise 3 must refer to the same kind of freedom as premise 2. If premise 2 refers to freedom from causation, then so must premise 3. But premise 3 is eminently questionable if it refers to freedom from causation. Then it claims that any agent whose act is not free from causation is also not responsible. That claim again begs the question against compatibilists. Because it is controversial, it needs to be supported by an independent argument, but no such independent argument has been given yet.

Thus, the contrastivist account of freedom shows that the argument from determinism commits the fallacy of equivocation. Of course, that is not the end of the discussion. Several responses are available.

Incompatibilists (including libertarians and hard determinists) might simply deny that they equivocate, because they refer to freedom from causation throughout, and premise 3 is true because all causes are excuses and do remove responsibility. Even if this is true, however, it is controversial and not at all obvious. Indeed, it conflicts with extremely widespread beliefs and practices, such as when juries believe criminals are responsible and hold them responsible. Moreover, it is a strong, abstract, universal, modal claim. Such premises cannot simply be taken for granted. For these reasons, the burden of proof lies with incompatibilists to show why all (or at least all deterministic) forms of causation always remove responsibility.

Most incompatibilists accept this burden of proof and provide support for their premises. One common argument runs like this:

2*. If any act is (fully) determined, then its agent cannot do otherwise.
2**. If any agent cannot do otherwise, then that agent is not free (at all).
2. Therefore, if any act is (fully) determined, then its agent is not free (at all).

This argument is supposed to refer to freedom from excuse in both premise 2** and conclusion 2. If so, this new argument would avoid the equivocation above. However, instead of escaping equivocation altogether, this argument merely introduces a new equivocation. To say that an agent cannot do otherwise might be to say either that the agent is prevented by some cause or that the agent is prevented by some condition that excuses him. Consider a student who says, "I can't go to the party tonight, because I have homework to do." This student is saying that he has an excuse (or justification) for not going to the party. He is not saying that he is prevented by some determining cause. But then what does "cannot do otherwise" mean in the argument above? If "the agent cannot do otherwise" refers to being prevented by some cause, then premise 2* is trivial and true by definition, but premise 2** is controversial and begs the question, assuming that it refers to freedom from excuse. On the other hand, if "the agent cannot do otherwise" refers to being prevented by some excusing condition, then premise 2** is trivial and true by definition, but premise 2* is controversial and begs the question. Hence, this argument again depends on an equivocation that is revealed by specifying contrast classes, just like the original argument that posed the problem of determinism.

Of course, incompatibilists have given many more arguments for the crucial premises, but these arguments all have forceful critics. I cannot go through all of these variations here (see Kane 2011). All I can do in this brief chapter is register my claim that none of these arguments succeeds completely, and many of their flaws are revealed by analyzing them in light of contrastivism along the lines illustrated above.

Apart from formal deductive arguments, the central question in this debate is why excuses excuse. We normally excuse people when their bodies cause damage if they were pushed or had a seizure. These kinds of causes excuse, but why? A natural answer is that the harmful movement was determined so that the person could not do otherwise. If that explains why these excuses remove responsibility, then that explanation would seem to generalize to all causes of action.

The problem with this inference to the best explanation is that this explanation fails in other cases and another explanation does better. First, consider coercion. When a robber credibly threatens, "Your money or your life," then you are not responsible for handing over the money even if you are free from causation and not determined. Because coercion can excuse *without* determining the agent's compliance, determination and freedom from causation are not what explains why this excuse excuses (or why this defense is a defense, if coercion counts as a justification rather than an excuse).

Similarly, reasonable mistakes can also excuse without determining. Suppose that you reasonably believe that the white granules in the sugar bowl are sugar, and your friend asks for sugar in her coffee, so you put some of the granules in her coffee. Unfortunately, the granules are poison, and she dies as a result. You are not responsible, and your excuse is complete, even if you are not determined to put anything in her coffee. Hence, determination cannot explain why this excuse excuses.

Again, imagine that an aggressor pushes you down onto an unseen stranger, and the impact breaks her leg. Suppose that the push was not so hard that you were totally incapable of resisting, but it would have been extremely difficult. If the aggressor had pushed you this hard a hundred times in similar circumstances, you would have fallen ninety-nine times; but once you would have managed somehow to avoid falling on the stranger's leg. Nonetheless, you do not seem responsible for the accident or the broken leg, because it would have been so difficult for you to achieve the contrary result. Examples like these suggest that determination cannot explain even the central cases of physical force that motivate incompatibilism.

What then does explain why excuses excuse? I would tentatively suggest roughly that excuses depend on what it is reasonable to expect of other people. It is not reasonable to expect someone to refuse to hand over money in response to a credible deadly threat. (Notice that it is reasonable to expect people to refuse to commit murder in response to a credible deadly threat, and the difference between murder and money has nothing to do with determinism.) It is also not reasonable to expect people to make mistakes that all or most other reasonable people would also make. (Notice that the difference between reasonable and unreasonable mistakes also has nothing to do with determinism, but does affect which mistakes excuse.) Finally, it does not seem reasonable to expect people to resist pushes (or compulsions or addictions) that are extremely difficult to resist, even if they are not completely irresistible. When we survey the range of actual excuses, then the real rationale for excuses concerns what it is *reasonable to expect*. That has nothing to do with determinism. This result leaves no basis for the premises needed for the argument that posed the problem of determinism. Because the burden is on those who assert those premises, the problem of determinism has been avoided with the help of contrastivism.

So far I have written as if the debate is between incompatibilists and contrastivists. This way of framing the discussion can mislead, because contrastivists are no more compatibilists than they are incompatibilists. Contrastivists hold that freedom from excuse is compatible with determinism, but they also hold that freedom from causation is incompatible with determinism, and both kinds of freedom are legitimate notions. Contrastivists are, then, both compatibilists (about freedom from excuse) and incompatibilists (about freedom from causation) as well as neither compatibilists (about freedom from causation) nor incompatibilists (about freedom from excuse). If someone asks a contrastivist whether he is a compatibilist or an incompatibilist, then he should reject this simple question and ask, "Compatibilist about what? Which kind of freedom are you asking about?"

Some critics get impatient and insist, "But which kind of freedom matters? In the end, are people really just plain free or not?" To say that some person or act is just plain free is, presumably, to say that they are free from the relevant constraints. The relevant constraints are then simply the ones, whichever they are, that determine whether a person or act is just plain free. This is obviously circular, but the point is that those who call a person or act just plain free without explicitly mentioning any set of constraints still presuppose that a certain set of constraints is the important or relevant one.

I refuse to play that language game. I see no point in arguing about whether any person or act is or is not just plain free without reference to any contrast class of constraints. That is because I see no basis for claiming that either freedom from cause or freedom from excuse is the one relevant contrast class for determining whether or not an agent is really and truly just plain free. Hence, I suspend belief about which contrast class is really relevant for freedom; about which acts and agents, if any, are just plain free; and about compatibilism or incompatibilism regarding unqualified kinds of freedom. This feature makes my position Pyrrhonian and enables it to avoid futile disputes that plague traditional discussions.

3   FREE TO DO WHAT? CONTRASTING OPTIONS

It is not enough to specify what a person is free *from*. We also need to cite contrast classes in order to specify what a person is free *to* do, as Feinberg said in the quotation above.

This new dimension of contrasts can be illustrated by Al the alcoholic. Al drinks heavily almost every evening but rarely at work. He knows that his drinking causes him personal and health problems, and he does not like or enjoy drinking any more, but he still wants to drink, and he would suffer withdrawal if he quit. He thinks a lot about drinking and spends a lot of time seeking drinks. In the lingo, Al is a heavy drinker, an abuser, dependent, and an addict.

Al needs money. If you offer him $20 to carry a ten-kilo bag of dog food for you, he will carry it for the money. But if you offer him $100 or $500 to carry a fifty-kilo bag of coal for you, then he will want to do it and try to do it, but he will fail. If you then credibly threaten to punch or kill him if he does not lift the fifty-kilo bag of coal, then he still won't do it. This shows that he cannot lift that much. He is not strong enough physically.

Similarly, Al's ability to control his drinking is shown by how he reacts to various incentives not to drink. If you offer Al $100 a day to stop drinking whisky, he will switch to wine and take the cash. If you offer Al $100 a day to stop drinking alcohol of any kind, then he will want to do so, he will try hard, and he will usually abstain all morning and afternoon and for a few hours into the evening. Despite great efforts, he will almost never make it the whole day without drinking any alcohol, so he won't succeed in collecting the money.

Is Al free? That question is hard to answer until we add contrast classes. Al is free to stop drinking whisky in contrast with wine, but he is not free to drink only soft drinks in contrast with alcohol. He is free to stop drinking for an hour as opposed to only thirty minutes, but he is not free to stop drinking for a day or a week in contrast with only an hour. We need to specify the relevant contrast class in order to describe precisely what Al is free to do and what he is not free to do.

The same point applies to individual acts. One day Al drinks a shot of whisky at 9:00 p.m. Was that particular act of drinking free? This act of drinking a shot of whisky at 9:00 was free in contrast with drinking wine at 9:00 and also in contrast with waiting until 9:15 or 10:00 to drink anything but not in contrast with drinking nothing but soft drinks until tomorrow. Again, we need to specify the relevant contrast class in order to describe precisely how an act is free and how free it is.

Indeed, much more precision is needed. For one thing, Al might respond more to stronger incentives. Suppose that for $1000 Al will stop drinking alcohol for one day but still not for two days. And suppose Al responds more to negative incentives than to positive incentives. If his boss threatens to fire him if he drinks on a weekend retreat, then Al will go without a drink for the weekend; but if the retreat lasts a whole week, then Al will try hard to abstain but will end up sneaking away for a drink before the week is over, despite the risk to his livelihood. And if Al takes Antabuse (Disulfiram), so he knows that he will get nauseous if he drinks alcohol, then he will usually abstain from drinking for a week, but still not for a month. He honestly tries hard to go longer without drinking, and he feels regret when he fails, but he does fail.

Circumstances also matter. Maybe for $100 Al will abstain for ten hours if he is at home with a supportive friend but not even for one hour if he is at a bar with others who are drinking. Personal tragedies might also lead him to drink within ten minutes even if he could save his job by waiting an hour.

Another complication is that, instead of always succeeding or always failing when they try, most people succeed at variable rates. Suppose Al expects no payment for not drinking and no sickness from drinking, but he does want to abstain, so he tries hard to abstain. Then he might often (70% of the time) succeed in abstaining for one hour, sometimes (20% of the time) abstain for two hours, but almost never (1% of the time) abstain for three hours.

I have described this case in some detail, because real cases are often at least this complex. Indeed, much more detail could be added. Addicts never have no control at all in any circumstances. Hence, we need to stop asking whether a person is free or in control and, instead, start asking how much control a person has or how free he and his acts are. The answer then needs to introduce contrasts, because most people are free to choose out of some contrast classes but not out of others.

But aren't some drinkers completely free from compulsion for any contrasting acts of drinking? Maybe, but their freedom still needs to be described in terms of contrast classes in order to specify the kinds of freedom that they share with alcoholics and the ways in which they differ from alcoholics.

Critics might become impatient and insist, "Stop it with all of the contrasts. Is Al just plain free? Is his act free period?" What do such questions mean? We already specified that Al is free out of some contrast classes but not others, so the questions seem to be about which contrast class is important or relevant. The problem is that different contrasts are relevant to different purposes. The kind of freedom that is necessary for blaming Al might be very different from the kind of freedom whose lack shows that Al needs his friends to stop offering him alcohol or drinking in front of him or that Al needs to start attending Alcoholics Anonymous. There is no way to determine whether a contrast is or is not relevant without picking a particular purpose and asking whether Al's kind of freedom is adequate for that purpose. However, a philosopher needs to avoid privileging one purpose over others if he wants a general theory that is neutral among purposes and merely describes the ways in which Al is and is not free.

As a result, I doubt that any contrast among options is the relevant one for determining whether or not an agent or an act is really and truly free. That doubt makes me a meta-skeptic about the real relevance of contrast classes.

To claim or deny that some agent or act is just plain free without mentioning any contrast class presupposes that some unmentioned contrast class is the relevant one. Hence, a meta-skeptic about real relevance should refuse to claim or deny that anyone or any act is just plain free. For this reason, I suspend belief about all such unqualified claims and denials. That makes me Pyrrhonian. When someone asks whether an act or agent is free or not, I refuse to answer directly and, instead, answer with a question: free to do what as opposed to what? Other philosophers should join me in this position or else show why one contrast class is the one that is really relevant.

4   INTERACTIONS

The two dimensions of contrast in freedom interact. Each constraint that I am free from is associated with its own contrast class of what I am free to do. Sometimes these classes overlap, but not always.

Suppose that a robber threatens, "Your money or your life." In addition to my wallet, I have a cell phone in my pocket. All the robber wants is money, but I also hand over my cell phone, because I have a compulsion to obey authority figures (or maybe my fear drives me to be unnecessarily cautious). In this case, how free am I? I am free from *coercion* to hand over my money but keep my cell phone in contrast with handing over both. But I am not free from *compulsion* (or fear) to hand over only my money in contrast with both my money and my cell phone.

Again, suppose that, while walking, I trip and stumble in such a way that I know I will fall soon, but I can manage to fall either to the left or to the right. To the left is a valuable vase, and to the right is a cheap imitation, and I will break the vase in the direction of my fall. I do not know which vase is valuable. Only experts could tell, and I am no expert. In that case, how free am I? I am not free from physical force (the force that tripped me) not to fall at all in contrast with falling. Nonetheless, I am free from physical force to fall either to the right or to the left, so I am also free from physical force to break the cheap imitation in contrast with the valuable vase. And yet I am not free from ignorance to break the cheap imitation rather than the valuable vase, because I do not know which is which, so my ignorance prevents me from making that choice.

It is difficult to combine all of these factors into a unified analysis of freedom of action. But here is a tentative rough suggestion:

A person, P, is free from constraints, C, to do an act, A, in contrast with options, O, if and only if both A and at least one alternative to A in O are not ruled out by any constraint in C.

This account surely needs refinements, but the basic idea is clear, I hope. Contrast classes C and O specify which constraints P is free from and which options P is free to do. Then to be free to do A, it is not enough that no constraints in C prevent P from doing A. After all, no constraints prevent a prisoner from staying in a locked jail cell, but he still does not stay there freely. In addition, there must also be no constraints in C that prevent P from doing otherwise—that is, from doing some other act in option set O. Different constraints in C might rule out different options in O, but P is free from C to do A out of O only if P has some choice in O other than A that is not ruled out by any member of C. Freedom requires an alternative.

5 DEGREES

One lesson of this contrastivism is that freedom comes in degrees. Some agents and some acts are more free than others. This point might seem obvious, but it is often overlooked.

One reason for this oversight might be that determinism and freedom from causation seem dichotomous. It seems that either an act is fully determined or it is not fully determined.

Reality is not so simple, however, because causes can limit the available options to a set without reducing options to only one single act (cf. Schaffer 2005b). An agent might be determined to do some act within a certain class in contrast with any act outside of that class, and yet not be determined to do any particular act within that class as opposed to other acts within that class. For example, even if Tony has freedom from causation to choose lasagna in contrast with manicotti (or vice versa), Tony still might be determined to choose pasta rather than curry if Tony is in an Italian restaurant that does not offer curry. Or Tony's love of lasagna might determine that Tony will order some kind of lasagna, but Tony still might have freedom from causation to choose vegetarian or meat lasagna, white or tomato-based lasagna, and so on. Determinists usually think that Tony is determined to choose one particular dish instead of any other, but it is possible that causation could limit available options to a contrast class without completely determining one particular act within that contrast class. If so, humans might have some freedom from causation but only within a limited contrast class. The classes of options can vary in size, and agents seem to have more freedom when they choose out of a larger class of options. Thus, even freedom from causation might come in degrees of a sort.

Nonetheless, almost all discussions of contra-causal freedom assume that freedom from causation is dichotomous: you've either got it or you don't, at least in a given case. After all, to say that an act was determined is to say that it was completely determined down to that act alone. There are no degrees of freedom on that view.

The picture is different when we turn to freedom from constraints that are more limited than causation in general. Freedom from excuse—including constraint, compulsion, ignorance, coercion, and prohibition—clearly comes in degrees, because each of these excuses allows degrees.

Consider coercion first. The degree of coercion varies with the amount of harm that is threatened as well as the probability of escaping harm if you don't comply with the demand. For example, if a small boy threatens to hit you if you don't give him your money, and if you can probably outrun him, then there is little coercion (or none if he is very small). But if that small boy has a gun, then running will create a risk of being shot, so the coercion is great. And if the boy has a knife or baseball bat but no gun, then running creates some risk of injury, because he might hit you before you get away or you might stumble. In this case, then, you are coerced less than if he has a gun and more than if he has no weapon. No reasonable person would succumb to the threat if the very small boy has no weapon, all reasonable people would hand over their money if the boy has a gun, and varying percentages of reasonable people would hand over their money if he has some other kind of weapon, depending on the kind of weapon, how fast he and they are, how much they need the money, and so on. Thus, some threats are very coercive, others are moderately coercive, and still others are only slightly coercive. Between the extremes lie many degrees of threat and coercion.

Similarly, only reasonable mistakes excuse, and the reasonableness of mistakes comes in degrees, because more or fewer people might be more or less likely to make the same mistake given a certain body of information. For example, if I cook a dinner with peanuts, but you are allergic to peanuts, so you get very sick, then whether my mistake was reasonable depends on how many people in the area are allergic to peanuts and how well-known peanut allergies are. Because these factors vary along a continuum, so does the reasonableness of my mistake. Some mistakes are totally reasonable, others are totally unreasonable, and many mistakes lie between these extremes.

Compulsions also vary in strength. Several variations were illustrated by Al the alcoholic above, but the same point also applies to kleptomania, agoraphobia, and other mental illnesses. Claustrophobes, for example, vary with respect to how large an elevator needs to be before they can enter it and also with respect to how long they can stay in each size of elevator when they are promised various positive and negative incentives to stay in it. Treatment often expands their freedom by increments.

Physical and legal constraints can also be more or less restrictive. A person who is locked in the trunk of a car has less freedom than a person who is confined to a prison cell, who in turn has less freedom than someone locked in his home or under house arrest. All of them have less freedom than someone who is not legally allowed to leave a certain country but may roam freely within that country. That person has less freedom than a normal citizen who is allowed to go abroad at will.

Because freedom comes in so many degrees, we often should not ask simply whether a certain agent or act is free from coercion, mistake, compulsion, physical force, or legal prohibition. Instead, we need to ask *how* free this agent or act is from each of these constraints. The answer will, of course, vary from occasion to occasion.

The same points apply to what an agent is free to do. Degrees can vary, first, with respect to the *type* of action: a person who can drink wine instead of whisky has more freedom than a person who cannot avoid drinking whisky. Degrees of freedom can also vary within a single type of action with respect to the *number* of acts: a person who can stop after two drinks has more freedom than a person who cannot stop before five drinks. What an agent is free to do can also vary with *time*: a person who can go a day without drinking has more freedom than a person who cannot go an hour without drinking. Another dimension of variation is *location* or *circumstances*: A person who can avoid drinking for an hour in a bar or at home has more freedom than a person who can avoid drinking for an hour at home but cannot go that long without a drink in a bar.

In general, people have more freedom when they are free with respect to more options. Additional degrees and contrasts arise because people vary in the probability that they will drink in the various cases above.

What has more or less freedom to act might seem to be agents rather than acts. Suppose that Al has to drink some alcohol within an hour, but nothing limits his choice between red wine and white wine other than his preference. His particular act of drinking red wine now rather than waiting an hour or choosing white wine instead of red might seem totally free from all constraints. He as an agent has more freedom than some and less than others, but this particular act seems free completely and not just free to a degree. This appearance is misleading, however. This act of drinking red wine at 9:00 p.m. is free in contrast with drinking red wine at 9:15 or in contrast with drinking white wine at 9:00. It is not free, however, in contrast with drinking only water or waiting until 11:00 for the first drink. This act is, therefore, less free in this way than a similar act done by an agent who could wait until 11:00 for the first drink. The comparative degrees of freedom of the agent, thus, lead to comparative degrees of freedom in particular acts as well.

6 APPLICATIONS

The best way to test a theory is often to apply it to cases to see whether it helps to illuminate or resolve issues. We can test contrastivism about freedom by applying it to an example that puzzled Aristotle:

> Those things, then, are thought involuntary which take place under compulsion or owing to ignorance. . . . But with regard to the things that are done from fear of greater evils or for some noble object (i.e. if a tyrant were to order one to do something base, having one's parent and children in his power, and if one did the action they were saved, but otherwise would be put to death) it may be debated whether such actions are involuntary or voluntary. Something of the sort happens also with regard to the throwing of goods overboard in a storm; for in the abstract no one throws goods away voluntarily, but on condition of its securing the safety of himself and his crew any sensible man does so. Such actions, then, are mixed, but are more like voluntary actions; for they are worthy of choice at the time when they are done, and the end of an action is relative to the occasion. (Aristotle 1941, 1110a)

Aristotle seems unsure whether to classify the act (presumably by the ship captain) of throwing the goods overboard as voluntary. Modern philosophers might be just as unsure whether to call the ship captain or his choice or act free.

Contrastivism clarifies this case. Although it is not clear whether to call the captain just plain free overall, it is much easier to determine whether the captain was free out of specified contrast classes. The captain *was* free from mistake and delusion, because he knew exactly what he was doing and why. The captain was also free from physical force, for it was not as if a wave hit him and made him drop the goods overboard by accident. He was also free from compulsion, because he had no volitional or emotional mental illness that led him to do what he did. And he was free from coercion by any other person, because the storm that created the danger was an impersonal force, not a robber. Still, the captain was *not* free from duress (or necessity, as some call it), because the circumstances created extreme dangers if he had not thrown the goods overboard. So he was free from some constraints but not others. We can tell where he stands on this dimension of freedom—freedom from—by specifying what he is free from and what he is not free from.

Next, what is he free to do? Here, again, we need to specify different contrast classes. The captain did *not* freely (choose to) throw cargo overboard as opposed to returning the cargo safely in his ship, for he never would have returned the cargo regardless of how much incentive he had and how hard he tried. Nonetheless, the captain did have other kinds of freedom to (where what he is free from is excuse). The captain *did* freely (will to) throw the cargo overboard as opposed to letting the ship sink. He also freely threw (and willed to throw) certain items rather than others (if he threw only part of the goods) or all instead of only part (if he threw it all). And, of course, he freely willed to throw the cargo at this time rather than somewhat earlier or later. The risk grew slowly, and he always could have waited and taken a little more risk. So he was free to do some things but not others, and he did what he did freely under some descriptions but not under others. This new way of thinking about the problem illuminates the issues and enables us to ask more precise questions about where this agent is located in the multidimensional space of possible degrees of freedom.

A more recent case where contrastivism about freedom helps is the trial of John Hinckley. Hinckley shot President Reagan in 1981, while Reagan was leaving a hotel. Hinckley's goal was reportedly partly to impress Jody Foster, the actress. Some of Hinckley's beliefs were true: shooting Reagan did get Foster's attention and did show her how much he loved her. Still, other of his beliefs were false: Hinckley's act would not create any real chance that Foster would reciprocate his love. So he was free from ignorance of some kinds but not others. In addition, Hinckley stalked Carter before Reagan, so there was reason to believe that Hinckley would have shot another politician instead of Reagan, if he had not shot Reagan. If so, he was free not to shoot Reagan, but he was not free not to shoot a politician (or at least he was not free not to shoot some famous person), because that was essential to his compulsive plan to impress Foster.

The issue of timing even made it into Hinckley's trial. Here is a bit of the transcript:

Q: [by Mr. Adelman, Prosecutor]. Let me ask you to focus on the moment when President Reagan leaves the limousine and walks into the hotel. Okay?

A: [by Dr. Carpenter, Defense Witness]. Yes.

Q: Mr. Hinckley was there with the gun, right?

A: Yes.

Q: And he could have shot him if he wanted to, right?

A: Yes.

Q: But he elected not to shoot him because he didn't have a good shot, right?

A: No, he did not act on that impulse at that moment.

Q: Well, if the impulse was overwhelming, why didn't he shoot him when he first saw him at 1:45 when he walked into the hotel?

A: It was the same sort of thing why he didn't shoot himself at the Dakota, why he didn't shoot Reagan early in December and why he didn't shoot Carter. This whole drive of balance and impulse and the thing that makes one hesitate—I think the ability to hesitate has become sharply eroded and I think that personalized experience as Mr. Reagan comes out of the limousine is a further erosion in that, but he did not pull the gun out and fire at that time.

Q: If the ability to hesitate was eroded when the President got out and waved, why didn't Mr. Hinckley with the ability to hesitate eroded shoot him then?

A: Because it is not one way or the other and this is a balancing of many factors and there is no way to give you a precise, emphatic answer to why he didn't shoot then, and shot when he came out. (Low, Jeffries, and Bonnie 1986, 73–74)

Here, prosecutors suggest that Hinckley must have been free not to shoot Reagan on the way out, because he did not shoot Reagan on the way in.

That argument impressed many people at the time, but contrastivism shows why it is specious. Just as a smoking addict can avoid smoking for a few minutes until he gets out of the building but cannot go a whole day without smoking at some point, so Hinckley might have been able to avoid shooting on a particular occasion or over a period of time, even if it was true that he would inevitably shoot some politician eventually and could not stop himself over the long run. The facts of the case are, of course, in dispute, but the point is that the observation that Hinckley did not shoot Reagan on the way into the hotel does not imply that Hinckley was in control or free to conform to law for longer periods of time. Contrasts help us to see through this fallacy.

The prosecutor is likely to retort, "Quit playing around. Answer the question: Was Hinckley free? Did he act freely?" (Or was he able to do otherwise? Or did he have the capacity to conform to law?) Contrastivists hold that such questions should be rejected and replaced by more precise questions with explicit contrast classes on at least the two dimensions discussed above: freedom from and freedom to.

A major advantage of adding contrast classes here is that, even if we disagree about whether someone is just plain free, we still might be able to agree that he is free to avoid misbehaving in a certain way rather than in another way or that he is free to avoid misbehaving at all for ten minutes but not for ten hours or ten days. This precise agreement can then guide our discussion of which kind and degree of freedom is necessary and sufficient for responsibility or for some other purpose. Contrastivists might still reach different conclusions on these issues, but the new questions at least make the issues more precise, enable agreement on some issues, and make it clear where the remaining disagreement lies.

7   CONCLUSION

My goal has been to show that contrastivism about freedom is coherent, plausible, and fruitful. Its main negative benefits are to avoid questions that are too imprecise to admit of definite answers and, thereby, to avoid needless disputes. Its main positive benefits are to point the way toward more precise questions that enable more agreement and illuminate what is and is not at stake. That would be progress. I do not claim to have made much progress here. All I hope is to have suggested some reasons why it might be useful to rethink old debates about freedom in terms of contrasts and degrees.[1]

NOTES

1. I am grateful for helpful comments from many members of audiences at Oxford University, Washington University in St. Louis, Duke University, Dartmouth College, and the Central States Philosophy Association.

REFERENCES

Aristotle. 1941. *Nicomachean Ethics*. Trans. W.D. Ross. In *The Basic Works of Aristotle*, ed. R. McKeon. New York: Random House.

Austin, J.L. 1961. *Philosophical Papers*. Oxford: Oxford University Press.

Craver, C. 2007. *Explaining the Brain*. New York: Oxford University Press.

Dretske, F. 1970. Epistemic Operators. *Journal of Philosophy* 67(24): 1015–1016.

Dretske, F. 1972. Contrastive Statements. *Philosophical Review* 81(4): 411–437.

Feinberg, J. 1973. The Idea of a Free Man. In *Rights, Justice, and the Bounds of Liberty: Essays in Social Philosophy*, 3–29. Princeton: Princeton University Press, 1980. Originally published in *Educational Judgments*, ed. J.F. Doyle, 104–124. London: Routledge & Kegan Paul, 1973.

Goodman, N. 1955. *Fact, Fiction, and Forecast*. Cambridge, MA: Harvard University Press.

Hume, D. 1739–1740. *A Treatise of Human Nature*. Ed. L.A. Selby-Bigge. Oxford: Clarendon Press. Clarendon Press edition published 1888.

Kane, R. 2011. *The Oxford Handbook of Free Will, Second Edition*. New York: Oxford University Press.

Karjalainen, A., and A. Morton. 2003. Contrastive Knowledge. *Philosophical Explorations* 6(2): 74–89.

Karjalainen, A., and A. Morton. 2008. Contrastivity and Indistinguishability. *Social Epistemology*, 22(3): 271–280.

Lipton, P. 1991. *Inference to the Best Explanation*. London: Routledge.

Low, P.W., J.C. Jeffries, and R.J. Bonnie. 1986. *The Trial of John W. Hinckley, Jr.: A Case Study in the Insanity Defense*. Mineola, NY: Foundation Press.

Schaffer, J. 2004a. From Contextualism to Contrastivism. *Philosophical Studies* 119: 73–103.

Schaffer, J. 2004b. Scepticism, Contextualism, and Discrimination. *Philosophy and Phenomenological Research* 69: 138–155.

Schaffer, J. 2005a. Contrastive Knowledge. In *Oxford Studies in Epistemology* 1, ed. T. Gendler and J. Hawthorne, 235–272. Oxford: Oxford University Press.

Schaffer, J. 2005b. Contrastive Causation. *Philosophical Review* 114: 327–358

Schaffer, J. 2008. The Contrast-Sensitivity of Knowledge Ascriptions. *Social Epistemology* 22(3): 235–245.

Sinnott-Armstrong, W. 1988. *Moral Dilemmas*. Oxford: Blackwell.

Sinnott-Armstrong, W. 2004. Classy Pyrrhonism. In *Pyrrhonian Skepticism*, ed. W. Sinnott-Armstrong, 188–207. New York: Oxford University Press.

Sinnott-Armstrong, W. 2006. *Moral Skepticisms*. New York: Oxford University Press.

Sinnott-Armstrong, W. 2008. A Contrastivist Manifesto. *Social Epistemology* 22(3): 257–270.

Unger, P. 1984. *Philosophical Relativity*. Minneapolis: University of Minnesota Press.

Van Fraassen, B. 1980. *The Scientific Image*. Oxford: Clarendon Press.

8 Luck and Fortune in Moral Evaluation

Julia Driver

Philosophers are interested in luck for a variety of reasons. One has to do with its purported significance with respect to moral evaluation and moral responsibility. Numerous philosophers have been fascinated and disturbed by the "paradox" of moral luck. We are, firstly, committed to the view that persons are only responsible for, or only blameworthy for, what they have control over. This condition is often referred to as "the control condition." It also seems to be a fairly obvious fact that we frequently don't have control over everything that happens as a result, for example, of our actions. Yet, those whose actions turn out worse than others who do exactly the same thing get blamed more harshly. Given the first two claims this does not seem warranted. The classic case is that of the reckless truck driver who has the bad luck to run over a child in the street. This truck driver is blamed far more severely than one who was equally reckless, but had the good luck not to run over anyone. Given that they both were acting equally recklessly the difference between the two is the result of luck, or chance. And thus the increased blame for the one who actually causes harm seems paradoxical—shouldn't they both be equally blameworthy if equally reckless?[1] This consideration is a major factor in pushing normative ethical theorists in the direction of purely internalist accounts of moral evaluation. On such accounts the moral quality of one's actions is completely determined by factors internal to agency, such as one's motives or intentions. Effects are irrelevant. Thus, what happens in the world as a result of one's actions is actually not a factor in moral evaluation of the action. On this view, both truck drivers are equally blameworthy in the sense that their actions are both equally *wrong*, equally reckless. On this approach the most common account of the phenomenon or paradox of moral luck is an *epistemic* account. What makes the truck driver case seem so paradoxical is that—given our limited epistemic resources—we can't genuinely tell if the two are equally blameworthy because we do not have access to their inner states.

The intuitive plausibility of this position presents the objective form of consequentialism, which offers an externalist account of moral evaluation, with a challenge. Objective consequentialism holds that consequentialism provides a criterion for evaluation. In the case of actions, that criterion is that the right action is the one that really does produce the best outcome. It does not hold—as the subjective consequentialist holds—that the right action is the one performed according to the consequentialist decision-procedure; or the one that, for example, maximizes expected rather than actual utility. Thus, for the objective consequentialist what actually happens as a result of one's actions determines its rightness or wrongness, although not, perhaps, its blameworthiness or praiseworthiness. Because factors external to agency—such as consequences—determine rightness or wrongness, the account is an externalist one for this species of moral evaluation. The challenge for this approach to moral evaluation is to account for moral luck without giving ground to the internalist. The overarching goal of this paper is to meet this challenge while, along the way, trying to clarify what the problem of moral luck consists in. Some of what comes under the heading of "moral luck" isn't actual luck, but, rather, good or bad moral fortune. The underlying problem of moral "luck" has to do with people getting credit or discredit for what they, intuitively, at any rate, don't deserve. This can arise through luck, that is, through fluke or accident. But it may also arise in nonaccidental ways.

A secondary goal of the paper is to attempt to arrive at a better understanding of luck itself. Few accounts of moral luck offer an account of luck itself. The view I argue for here holds that the best account of luck itself is *contrastive*. This means, among other things, that no one is just plain lucky or unlucky. We have various pragmatic rationales for identifying an outcome as lucky or unlucky. In the case of morality, I will maintain, some of the relevant reasons have to do with what it is reasonable to blame and praise someone for. Outcomes will be relevant here because we want to minimize the actual bad results of actions. Bad "willings" are to be reduced because these are what have a causal connection to the bad outcomes. The type of luck that I will primarily be concerned with in this paper is resultant luck, or luck in consequences.

1   THE PROBLEM OF MORAL LUCK

The general problem of moral luck is by now well known. Discussion of the problem in the contemporary literature was stimulated by articles by Bernard Williams and Thomas Nagel (see Nagel 1979; Williams 1981). Nagel, for example, notes in his presentation of the problem that if we succeed or fail in our projects it is often a matter of luck. There are some things, indeed, plenty of things, we simply have no control over. He writes:

> However jewel-like the good will may be in its own right, there is a morally significant difference between rescuing someone from a burning building and dropping him from a twelfth-storey window while trying to rescue him. Similarly, there is a morally significant difference between reckless driving and manslaughter. But whether a reckless

driver hits a pedestrian depends on the presence of the pedestrian at the point where he recklessly passes a red light. (1979, 25)

A Kantian might well agree with Nagel's claim in this paragraph, but not view this as a problem for moral worth. Indeed, the Kantian system is actually constructed so as to avoid the impact of moral luck on moral worth. It is a theoretical strength of the Kantian position that it insulates moral worth from luck. Thus the challenge for externalist accounts. It is also important to note that one could adopt an internalist stance with respect to moral evaluation and not be committed to a Kantian position. Indeed, subjective consequentialists are internalists for pretty much the same reason Kantians are—to avoid moral luck. Subjective consequentialists hold that the moral quality of one's action is determined by the subjective states of the moral agent that are internal to agency. So, on one popular construal of this approach, the right action would be the one that maximizes expected utility, where expected utility is understood as what the agent expects to maximize utility.[2] Usually, this view also builds in some kind of reasonableness requirement on the agent's expectations. Unlike the Kantian, however, the subjective consequentialist considers effects in practical deliberation. Thus, for the subjective consequentialist one is shooting for success in terms of outcome. It's just that these outcomes aren't relevant in measuring the success of the agent or the agent's action in moral terms.

The intuition elicited in the above case seems to be that if the reckless driver truly had no control over the presence of a person in the road, if the presence of the person was actually due to bad luck, he does not deserve extra blame for running over that person. We will come back to this, because I believe that the objective consequentialist can account for this intuition, properly construed.

If we just look at what we ordinarily tend to think about right and wrong it looks like we hold people responsible both for the moral quality of their mental states—their intentions and motives, etc., as well as for the outcomes of their actions when those outcomes are thought to have been guided by the agent's psychological states. We also blame persons and hold them responsible for outcomes in cases where they—although not guided by bad psychological states—are acting in the *absence* of the appropriate psychological states. Even if the agent didn't know better there are often situations where she should have.

An agent's motives, intentions, and so forth are thought to indicate what sorts of reasons the agent is responsive to. One can tell if the agent is moved by morally good reasons by looking at what she intends to do, or the sorts of motives she has in acting a certain way. Likewise, outcomes tell us the agent's actual impact on the world, something many people intuitively also think is morally relevant and morally important. Thus, most people very often have a kind of mixed view when it comes to morally evaluating what someone does. On this mixed view, we have to look at both the character of the agent's psychological states as well as the actual impact that her behavior has on the world. Further, the mixed view is committed to holding each of these factors—that is, the states internal to agency as well as outcomes—as somehow intrinsically important to the evaluation. That is, on the mixed view, neither can be given a reductive analysis in terms of the other. I don't think that this lovely compromise view withstands scrutiny, although I do think it represents our unreflective views on moral evaluation.[3]

The importance of intention to evaluation is taken to be demonstrated by the fact that we blame agents for intended bad outcomes far more severely than unintended ones. For example, if Samantha intentionally strikes Beatrice, Beatrice will be far more angry and resentful than if Samantha unintentionally strikes her while waving her hand. This is because, at least in part, Beatrice will understand the intention to harm her as far more threatening, and certainly indicative of the fact that Samantha views harming Beatrice as a reason for performing that action. Not so in the unintentional case.

When it comes to the significance of outcomes people will frequently note that the agent's impact on the world is morally significant—and to deny that significance encourages a kind of moral solipsism. Indeed, this is a major problem for internalist accounts of evaluation. To counter this problem they need to build into the theory substantive assumptions to the effect that morally good people just are the sorts of people who display a concern for what happens in the world.[4] This appears to be a backdoor acknowledgment of the significance of objective factors.

But some would argue that it is not morally appropriate to factor in what would happen were one not to perform the action. This is because some actions are intrinsically wrong, blameworthy, and ought not be done, even if they are instrumentally *good*, let alone outcome neutral. The mistake of consequentialism is to equate instrumental goodness with rightness and praiseworthiness. An action can be wrong even if, instrumentally, it is the "best" (in terms of producing the most good). Thus, when it comes to practical deliberation such actions should not be considered at all. This position has well-known problems and as stated does not withstand reflection. Ardent deontologists often note that when the consequences are good enough it will be permissible, even obligatory, to perform actions that are normally immoral. So, if we think of "intrinsically wrong" as "wrong in all contexts" then such actions are not really intrinsically wrong. They may be *prima facie* wrong, that is, they appear wrong on first blush, but the consequentialist could readily agree with that. Then the debate centers on where to draw the line on how bad things have to get to reach the "wrong" threshold.

But my aim here isn't to go over this debate between consequentialists and non-consequentialists—it is simply to point out that very many people do intuitively think that the agent's actual impact on the world is *morally* relevant, not simply causally relevant.

The moral luck problem arises out of our confused reactions to factors that each seem important to moral evaluation. Outcomes matter, luck ought not to matter, and yet luck and outcomes often go together. The externalist strategy will be to stick to her position on outcomes, at least in terms of assessing the agent's success in performing right actions, yet offer an account of moral evaluation that is nuanced enough to accept luck without serious violence to our moral convictions. We evaluate more than the rightness or wrongness of actions. We also evaluate persons themselves as praiseworthy or blameworthy. We evaluate the mental states of agents. A person may act rightly in such a way as to reflect badly on her character; or she may act wrongly in such a way as to reflect well on her character. Rightness and wrongness are subject to resultant luck, then. To better understand this strategy it will be useful to have some idea of what we are talking about when we discuss moral *luck*.

2   LUCK

No one is just plain lucky. In fact, one's luck status may seem murky in any given situation. Consider the following example:

> Sandra has had a narrow escape. She contracted an extremely rare, and extremely fatal, strain of flu. Fortunately, however, after two weeks of agonized suffering she has recovered and is recuperating in the hospital. Furthermore, through some odd and highly improbable combination of chemical factors the flu seems to have cured her arthritis. When her brother Bob comes to visit her she tells him happily: "I am so lucky!" Bob disagrees with her, claiming that in reality she has been quite unlucky.

Both Sandra and Bob are correct. And this example illustrates two distinct ways in which Sandra is both lucky and unlucky. Sandra is lucky to have caught the flu and then recovered, rather than died. She is also unlucky to have caught the flu, rather than to have avoided it altogether. A careful reader might note that this is not a genuine case of contrastivism—because, although Sandra is lucky to have caught the flu *and then recovered*, it would be odd to say that she is lucky to have caught the flu period. This lucky/unlucky contrast is not controversial at all—it simply reflects the fact that what one is lucky (or unlucky) *about* in a situation can vary within the situation itself. The *contrastive* luck attributions are the following: Sandra is unlucky to have caught the flu rather than to have avoided the nearly fatal disease. Sandra is lucky to have caught the flu, rather than to have continued to suffer from her arthritis. I will discuss further the issue of what makes the contrastivist approach distinctive later in the essay.

This scenario illustrates the contrastive nature of luck attributions that I'd like to explore in this paper. It also illustrates one of the conditions under which warranted attributions of luck, and lack of it, can be made.

Nicholas Rescher notes that in many cases lucky (or unlucky) means something like "by accident" or "by chance" (Rescher 1995). That is, the lucky (or unlucky) event was unplanned, or it was something that the agent could not have reasonably expected to occur (ibid..). Further, the event has normative significance "in representing a good or bad result, a benefit or loss" (Rescher 1990, 7). But this doesn't exhaust how we understand luck, either. We also associate luck with what is improbable: good luck would be the improbable with positive normative significance, bad luck the improbable with negative normative significance. I can plan or intend to climb Mt. Everest, while realizing that my odds of getting to the top are pretty low, and this would be sufficient to warrant a judgment that I was "lucky" to get to the top.

Further, as the moral cases have demonstrated, we also think of lucky or unlucky outcomes as those beyond the agent's control. Although the truck drivers can control their states of recklessness, they cannot control whether or not a child runs into the street. The morally unlucky truck driver is the one driving recklessly when the child runs into the street. As other writers have noted, however, lack of control can *at best* be necessary—it is certainly not sufficient.[5]

It will turn out that our attributions depend on pragmatic features, some of which pick out certain contrasts. In the case of morality, some of those pragmatic features will relate to our interests in blaming and praising people. Although it seems pointless to blame people or hold them responsible for things they had no control over, if one adopts a genuinely instrumental account of praise and blame then justifications for the practice can expand to include third part effects—thus, there may indeed be some small point in the blame, under certain restrictive circumstances.

The basic account of "luck" attributions that seems correct to me is something along the lines of:

> (CL) Event e is lucky or unlucky for a given individual in contrast to some other state of affairs (or, *rather than* some other state of affairs).[6]
> An individual, S, is lucky that p rather than q.

In his work in epistemology, Jonathan Schaffer argues that knowledge is contrastive.[7] It consists in a three place relation, Kspq (s knows that p rather than q), q being the contrast proposition. But an account can be contrastive and admit of further variables.[8] On the account of luck that I will be presenting, a fourth place will be called for because "luck" will be relative to the agent's interests as well. Thus, (CL) will need to be modified so that an individual s is lucky that p rather than q, relative to her set of interests and, in some cases, her epistemic states. It may well be that contextualists would argue that the account I present is really a contextualist one. However, I don't want to get involved in that debate here. On either way of presenting the account, luck attributions are dependent on relevant contrasts being specified, whether or not we simply hold those contrasts to be a matter of context.

Thus, for an event to be considered either lucky or unlucky that evaluation is relative to the judger's epistemic status as well as the judger's normative commitments. I leave aside for now the question of whether luck is more objective. Just as a man can be both tall and not tall, an event can represent both good luck and bad luck, depending on the perspective of the judger.

But it is first important to get clear on what is distinctive about the contrastivist approach. It is uncontroversial that judgments of luck are relative in various ways. They can be relative, for example, to the person, relative to the interests of the person, and/or relative to the circumstances of the person. For example, it is lucky for me if my enemy trips in battle, but unlucky for him. This shows that judgments of luck are made relative to the person. I might even remark, "Wow, lucky for me—but not for him!" This also underlies the comparative luck judgments: A is luckier than B with respect to e. I am luckier than my enemy with respect to our aims (his of killing me, mine of staying alive).

But the contrastivist is claiming more than this. The contrastivist is claiming that luck attributions—even with respect to the same person, the same set of interests, and the same circumstances, even holding all of these constant—are subject to contrasts. The Sandra case we began the section with illustrates this. Another case is the following: suppose that Roger's grandfather has just died and left him ten million dollars in his will, on the condition that Roger is not already a millionaire. Suppose also that Roger has just won the lottery for one million dollars. Roger is both lucky and unlucky. He is lucky to have won the lottery, rather than to have lost, given the improbability of winning. On the other hand, he is unlucky given the contrast with the contents of his grandfather's will. That is, he is unlucky to have won the lottery, rather than to have qualified for the ten million dollars in his grandfather's will. Articulating a contrastivist account of luck helps to clarify reasons why we judge someone or some outcome lucky or unlucky. Reasons themselves are understood relative to contrasts. The reason Sandra is lucky is that she caught the flu that eradicated her arthritis; she is lucky to have caught the flu, rather than something else, let's say, that would have had no impact on her arthritis.

Some people use the word "luck" to indicate a state that "could have been otherwise." It is just a matter of "luck" that one's parents happened to meet when they did, for example, because it could have been otherwise. But this needs to be narrowed a bit too, because utterances like "it is just a matter of luck that I am wearing my blue jeans today (because it could have been otherwise)" are absurd in normal situations. That's because I presumably *chose* to wear my blue jeans and there was nothing standing in the way of my choice. So, even though it is true that it could have been otherwise, it is not a matter of luck. If we deny this then the only things that aren't a matter of luck are those that are necessary. Although it is true that it is not a matter of luck that 2 + 2 = 4, that is not the whole story either. Control is a factor as well. That someone chose to do a and that a had the expected outcome (or, perhaps, the reasonably expected outcome) is a factor. Of course, luck judgments can apply to one's choice itself as well, if one thinks that one doesn't choose what to choose, or choose what to choose what to choose, and so on. A strategy to deal with resultant luck or circumstantial luck may not deal with this higher-level luck, related to constitutive luck.

Is there a way to take these inchoate impressions about luck and turn them into a more systematic account? My impression is that in the literature there are two main approaches to this issue. One I term *epistemic reductionism* because this view—which can be spelled out in a wide variety of ways— basically maintains that luck simply reflects a state of ignorance on the part of either the luck attributor or the "lucky" individual. The other view is more objective. It is called the *modal* view because it holds that luck is not simply epistemic but instead corresponds to flukes—occurrences of this world that fail to be occurrences in the relevant set of nearby possible worlds.

3   EPISTEMIC REDUCTIONISM

Again, one possible account is that luck—with respect to results or consequences—is essentially epistemic. That is, we would never think ourselves either lucky or unlucky if we knew all the facts. So, if I roll a pair of die and, hoping for the highest number, get two sixes I would likely consider myself lucky because I could not have reasonably expected that outcome given what I know and because that outcome is good for me given what I wanted. However, if I were in possession of full information then I would have known that throwing the die a certain way would result in double sixes, and thus the good outcome is not attributed to luck. Uncertainty is eradicated with full information.

This account of luck can be extracted from the work of Pierre Laplace. Laplace believed that a God-like being possessed of full information would not make judgments of luck.[9] For such a being there is no uncertainty about outcomes at all. On this account luck judgments are simply a reflection of our impoverished state of knowledge about what will happen and/or what is, in fact, good or bad for us. A God (or a "Laplacian demon") would not attribute luck to anyone because a God would have access to all the information—information about what has happened, the laws of nature, and in virtue of these two, what *will* happen as well. Added to this, we can suppose that God also has knowledge of what is good or bad for a person's interests. God makes no warranted luck attributions.

Nicholas Rescher also seems to hold a kind of epistemic reductionism. On his view judgments of luck are a matter of what the agent can reasonably expect to occur. Because one lacks full information there will be uncertainty and this provides the basis for luck judgments. Thus, a person who unknowingly benefits from a rigged lottery on Rescher's view is lucky because, from his point of view, he had no reason to expect that outcome. From his point of view, lacking the relevant information, the outcome was quite improbable.[10]

We can be ignorant of either the consequences of our actions or failures to act, as well as ignorant of whether or not those outcomes affect our goals positively or negatively. To give a slightly different case, George Bailey knows full well that if he doesn't leave Bedford Falls he will not get to be an engineer. What he doesn't know is that it might actually be a good thing for him that this doesn't happen. Thus, he might reflect back on his life and conclude that he was in fact lucky that he wasn't able to leave and become an engineer. This kind of ignorance involves ignorance of what is, in fact, normatively significant. Epistemic reductionism holds that our judgments of good or bad luck can be reduced to either this kind of ignorance or ignorance of what will, in fact, *happen*, or what is, in fact the case. Roughly, we can put the claim this way:

> (EpR) "A is lucky that e rather than f" is simply shorthand for "Given what the speaker knows about the likelihood of e's occurrence given A's circumstances, it was unlikely that e (or unplanned or uncontrolled, etc.) but not that f, and/or A did not know that e was in fact good for A, rather than f."

Thus, one would not judge A lucky that e if (i) one knew that e would occur and/or possibly if (ii) one was fully aware that e was a good thing for A. We want to be able to preserve the sense of claims like "Alan is incredibly lucky, he just doesn't know it." Also, however, we do tend to think that someone can be lucky or unlucky, and no one ever realize it or be in a position to realize it. For example, it is possible that we are all lucky that the planet Earth has not been struck by a giant meteor in the past two thousand years (as opposed to being struck by one), although we may never know this. What would the epistemic reductionist say about this? On this account there is no luck, *tout court*. We are not, in fact, lucky because it was inevitable that no asteroids would hit the Earth. Of course, if we actually thought about it given our limited knowledge of the universe we might well be warranted in thinking the probability of an impact high, and thus we are warranted in a judgment of luck. If, on the other hand, we had God-like knowledge of how the universe works—not actually accessible to any human being, of course—then we would not have made that judgment because we would know that there's no chance at all of the asteroid hitting the earth during that time.

There are a plethora of ways one could go about spelling out (EpR) in more detail. We cannot go into them all here, but I will mention two. First, we could hold that

> (EpR1) Attributions of "lucky" or "unlucky" are true or false relative to the epistemic states of the attributor.

Thus, when Sandra judges herself to have been lucky in catching flu, this claim is true given that she believes catching it to have been very unlikely. Of course, there is also no deeper issue here, no metaphysics of luck. Given full information she would not be warranted in attributing luck to her recovery.

We could also go the following route:

> (EpR2) Attributions of "lucky" or '"unlucky" are true or false relative to the epistemic states of the well-informed or reasonably well-informed attributor.

Suppose that Sandra's physician Nora knows that Sandra had been living in a flu "hot spot," although Sandra herself does not know this. Whereas on (EpR1) Sandra's judgment that she has been lucky is true, it is not necessarily true on (EpR2)—depending on how stringently one understands "well-informed." For example, Nora might well think luck had nothing to do with Sandra's contracting the flu—that, even though Sandra thinks it did, she's *wrong* about that.

It can get yet more complicated. In judging x to have been lucky ought one consider what the attributor believes or what the agent believes? In (EpR1) and (EpR2) I've spelled it out in terms of the attributor's epistemic states. But there are cases where the attributions seem appropriate relative to the agent's epistemic states *rather than* the attributor's. Suppose that Priscilla owns a store that sells lottery tickets and has just heard that the winning lottery number is #637845. Bill comes into the store at the last minute before the ticket sales are suspended and buys a ticket with that very number. Priscilla knows that there was no way for him to have known the number ahead of time. Under these circumstances she would be warranted in judging him lucky—but that makes sense only relative to *his* epistemic states. So neither (EpR1) or (EpR2) can model this type of case. We could try to modify them to read something like "relative to the epistemic states of the attributor and/or agent." However, as we will see shortly, there is a more streamlined way to proceed that will hopefully avoid this particular *ad hoc* maneuver.

It should be noted that there is also a moralized sense of "lucky." Suppose that Alice becomes engaged to Bob, her boyfriend for a number of years. Suppose also that this was no surprise to anyone. Alice might still sensibly utter "I am so lucky." Here she might mean something like "I have a great fiancé, much better than other fiancés, even though I am no more deserving of him than many other women would be." The idea is that we might judge someone lucky when something great happens to him or her— even if it is entirely predictable—as long as it seems that they are no more deserving than others of the great outcome. This counts against (EpR).

However, the epistemic reductionist could respond by holding that Alice's situation is not truly a matter of luck. Perhaps, for example, we acquiesce to such utterances out of respect for conversational norms. It may just be rude to contradict someone about her romantic good luck. The more serious issue for (EpR) is that it is fairly messy, because it mixes up various conditions that underlie luck judgments—such as the likelihood of the event, whether or not it was planned or intended, and whether or not the agent was exerting control. For example, several other writers have pointed out that it can't simply be lack of control that characterizes luck, because there are plenty of things we lack control over but are not lucky. One example of this in the literature is the following: we have no control over the rising of the sun each morning, yet it would be odd to say that this was a lucky occurrence.[11] Of course, one could come back and note that the rising of the sun each morning is highly probable, and that's why describing it as a lucky occurrence is odd. Pritchard argues that this can't underlie luck judgments, however, because, for example, a landslide's occurrence may be a matter of chance—although not described as lucky or unlucky unless it also affects someone's interests.[12] However, (EpR) adds a condition that would get around this particular concern.

Might there be a sense of "A is lucky that p rather than q" that is meaningful (true or false) independent of the epistemic states of an utterer, or even the agent? The issue is clouded by the fact that luck is tied to interests. Consider the following case:

> John rushes to the train station but, unfortunately, the train happens to be a bit early that day and he misses it. However, while waiting for the next train he happens to meet Lucy. Eventually, John and Lucy get married and live happily ever after. John, however, has forgotten by that time that he met Lucy as a result of missing the train. No one else is aware of that fact.

John, of course, believed himself to have been unlucky at the time he missed his train. If the train had been on time, as usual, he would have made it. Yet, it turns out, relative to his long-term interests it was actually lucky for him that he missed the train.

The epistemic reductionist is in a bit of a bind with cases like this. To avoid the rather counterintuitive result that John, in fact, has not been lucky, the epistemic reductionist needs to idealize a bit. But if she idealizes too far, then there is no such thing as luck at all. So there must be some sort of intermediate idealization. In the above case the epistemic reductionist might hold that John is lucky because, given what he should have been aware of at the time he and Lucy were married, the luck judgment is true. Or, she could hold that John is lucky relative to what a well-informed, although not actual, attributor would judge to be the case. The balancing act that the epistemic reductionist is called upon to perform in order to accommodate such cases is not impossible. But it will require a fluctuating standard regarding the relevant epistemic states.

Thus, is there an account of luck that can accommodate a *metaphysics* of luck? Again, on such a view it would be possible to characterize luck without appeal to the agent's or non-actual attributor's epistemic states. A contrastivist account divorced from epistemic reducibility can provide this. John is lucky that the train was early rather than on time, regardless of what he believes or what any actual attributor believes. Further, we needn't bring the non-actual well-informed (but not perfectly well-informed) attributor.

But note luck would still be understood *relative to a set of interests*. A rock is not lucky or unlucky, although its fate is subject to chance as much as a person's or an animal's. We can consign this normative element to pragmatics. That is, our interests, our purposes, or what is good for us—these are features of the situation that will make certain factors relevant in the attributions of luck. In the case of moral luck, however, it will turn out that this gets rather complicated, because we will want to consider not necessarily what a person's interests *are*, but instead what they *ought* to be. It is possible for a truck driver not to care whether or not he runs over anybody—of course such a truck driver would be evil—and it still is the case that he *ought* to care even if he does not. And moral luck is about evaluation of a person or a person's actions and his or her degree of moral responsibility, so whether he actually cares or not is irrelevant to ascriptions of moral luck.

4 THE MODAL ACCOUNT

One writer who has spelled out an objective account is Duncan Pritchard in his book on epistemic luck.[13] Pritchard argues that the best account of luck is modal, and consists of two conditions:

> (L1) If an event is lucky, then it is an event that occurs in the actual world but which does not occur in a wide class of the nearest possible worlds where the relevant initial conditions for that event are the same as in the actual world. (Pritchard 2005, 128)

This condition is not sufficient and he adds:

> (L2) If an event is lucky, then it is an event that is significant to the agent concerned (or would be significant, were the agent to be availed of the relevant facts). (132)

Pritchard argues that (L1) and (L2) combined are "clearly able to accommodate a number of our basic intuitions about luck" (Pritchard 2005, 133), although he also admits it is a rather vague account. (L1) is intended to capture the intuitions that lucky events are improbable, unplanned, accidental, and/or beyond the control of the agent. (L2) is supposed to capture the subjective nature of luck attributions. Without (L2) we'd get into oddities such as it is lucky for me that there are an odd number of stars rather than an even number of stars, even though I could care less about it.

A modal account will have issues with necessary truths. Consider the claim "I am lucky to have the parents I have." Pritchard would have to deny this, because it is necessary that I have the parents I have—I could have no other parents—there is no possible world in which I have different parents. Pritchard actually holds that his account is restricted to the nonnecessary. However, Pritchard could say that sometimes we misuse "lucky" to mean "fortunate."[14] I am indeed fortunate in my parents, but not lucky. It is good for me that I have the parents that I do have, although it was inevitable and thus not a matter of luck. It would be possible to handle the Alice case similarly. Alice is fortunate in her fiancé, but not truly lucky.

There are other issues for Pritchard's account. For example, the lucky event does not occur in "a wide class of the nearest possible worlds." But consider the following case: Michael is a very, very poor shot. Every day he goes out to the firing range to practice and, for the most part, performs miserably. He fails to hit the bull's-eye ninety-nine out of one hundred times. But each time he shoots he aims carefully, and clearly intends and wants to hit the bull's-eye. Then, on Thursday morning, he does hit the bull's-eye. Was he lucky? It would be hard to answer this given the account Pritchard puts forward. Whether or not Michael is lucky to have hit the target, rather than to have hit the edge of the target or something else altogether depends on the contrast we take to be operative.

Consider another example: A lottery has been rigged by Joe's father, Carl, so that Joe will win. The winning number has been picked ahead of time and Joe has been told what it is. Joe buys a ticket with the winning number. Carl also likes Sam, his best friend's nephew. He knows that Sam always plays the lottery and always picks his birth date as the winning number (060682). So, Carl picks that number when he rigs the lottery. Sam, however, is unaware of this. Lucretia also buys a lottery ticket that week, with the winning number. She has no connection to Carl and she, too, is unaware that the lottery has been rigged. Who is lucky? The contrastivist holds that there is no answer to this independent of a contrast. Sam is lucky relative to what he knew because he had no way of foreseeing the winning number. The same holds for Lucretia.

At issue, I think, is how Pritchard would unpack "the relevant initial conditions" in his account. What are these? How is "relevance" determined? Again, there are different ways we could go here. One is to insert an epistemic understanding of "relevant" and hold that the relevant conditions are the "foreseeable" ones—or something along those lines. Or, more radically, one could simply say that there are no truly relevant conditions. One just picks a class as the contrast class and makes the judgments relative to that class. For my purposes in this paper, however, I will be opting for the former because that ties quite naturally into the issue of what we find blameworthy and praiseworthy in agents.

If we develop the modal account in a contrastivist direction we get something like:

(CL1) If an agent is lucky that event p rather than event q, then p occurs in the actual world, and does not occur in a wide class of the nearest possible worlds where the relevant initial conditions for that event are the same as in the actual world, whereas q does not occur in the actual world, and does occur in a wide class of the nearest possible worlds where the relevant initial conditions for that event are the same as in the actual world; further

(CL2) Whether or not p constitutes good luck or bad luck is relative to the interests of the agent (or the being with interests).

Again, I would suggest that we combine the intuitive appeal of the epistemic approach with the modal approach by unpacking "relevant" in epistemic terms. The set of conditions that determines what is relevant are those that are the foreseeable outcomes. Foreseeable outcomes are those that could reasonably be expected to be foreseen—either by the agent, or attributor. It is quite true that this will sometimes not be clear. For example, in the case of the poor target shooter that I discussed earlier, this is not clear. Given the improbability of his hitting the mark one might argue that it was not foreseeable; on the other hand, he fully planned and intended to hit the mark, so, again, one might hold that it is foreseeable insofar as he planned to do it. But this simply demonstrates the relative nature of these judgments, and the meaning will be clear when the relevant contrast is made clear.

Sandra is lucky to have caught the flu rather than to have suffered with arthritis for the rest of her life. Sandra is unlucky to have caught the flu, rather than to have avoided the deadly disease. In both cases the flu constitutes an event that, due to its improbability, was not reasonably foreseeable. In the *relevant* class of nearby possibly worlds she does not catch the flu.

In Sandra's case we are assuming as part of the background that she has an interest in being healthy. That is her *actual* interest. In moral luck cases, luck is understood instead relative to the interests the agent *should* have. The attempted murderer whose gunshot is foiled by an improvident gust of wind, or an unlucky bird, is himself morally lucky, although not lucky in the purely descriptive sense. He is morally lucky to have shot the bird, rather than the intended victim, that is.

One ought not to be blamed (or as severely blamed) for outcomes of one's actions in this world when they did not occur in a wide class of the nearest possible worlds even when these outcomes are contrary to the interests one ought to have. If the truck driver ran over someone through sheer fluke—so that it is true that in nearby possible worlds he did not run over anybody, even though reckless, then he is no *more* blameworthy than the other, luckier, truck driver who did not run over anyone in the actual world. However, he still did something wrong that the other truck driver did not do, namely, run over someone. His action is wrong, and due to blameworthy recklessness, recklessness that is itself blameworthy for both of the agents. But the difference in intuitive reaction is due to the quite sensible observation that the lucky truck driver didn't do anything wrong *beyond* displaying recklessness which, by creating risk, endangered others even though it did not actually lead to a harm in this particular instance.

The morally unlucky truck driver is unlucky because he hits someone in the actual world, rather than merely speeding down the road without incident (as he intends); although, in the nearest possible worlds (with the relevant conditions fixed, etc.) he does not hit anyone. In the actual world he's done something *wrong*. As in the case of the other truck driver, we can read off from this his failure to properly acknowledge legitimate reasons for minimizing risk—reasons of safety—and this reflects quite badly on his character *as well*.

In the case of the attempted murder, the murderer is morally lucky because he hits a bird rather than his intended victim. He's done nothing wrong beyond the attempt. But again, this attempt is something that speaks badly of his character. In some nearby possible worlds (with the relevant conditions fixed, etc.) he has killed his intended victim. Again, the contrast—"rather than his intended victim"—demonstrates that he is not responding to the right sorts of reasons—he fails to value human life sufficiently. What is foreseeable, given the intentions, is the death of the intended victim.

This is why contrastivism is important to providing an insight into moral luck. The contrast can sometimes be designated by reference to the agent's intentions—or, more broadly, what the agent can reasonably foresee. It is these intentions that provide information about the sorts of reasons the agent takes seriously, or fails to take seriously, and these in turn provide insight into the agent's character and intentions. They give us information relevant to praise and blame of the agent, and what the agent foresees, reasonably, as an outcome of his behavior. This can provide the basis for luck judgments either considering simply what the agent foresees, or what the attributor foresees on behalf of the agent (given the agent's states of mind, etc.).

However, does this really solve the problem of moral luck? Someone might note that it handles cases where rightness/wrongness depends on outcomes, although blame does not. Blame depends on something else, the agent's states of mind and whether or not these reliably produce good. And this offers luck another foothold. Suppose we hold that an agent is blameworthy to the extent that he performs an action that he foresees will have an overall bad outcome. What he foresees is due to factors beyond his control; what he foresees may involve luck, or fortune. He controls his action based on what he foresees, true—but what he foresees itself is subject to "luck."

Whether or not this account is taken to solve or dissolve the moral luck problem will depend on whether or not one finds the problem at that level still troubling. We care about what agents intend, what they foresee, and what they can reasonably foresee because we rightly judge these to be factors guiding their actions as well as factors which indicate the sorts of reasons they are responsive to. Thus, this account can handle the problem with resultant luck. But luck with respect to one's epistemic situation is not handled. However, in the case of moral luck this is not something that strikes us as deeply problematic with respect to praise and blame. If someone could not have foreseen an outcome then this is taken to be relevant.

But not all cases of what is commonly termed "moral luck" are amenable to this analysis. The above characterization can handle our views of luck when it comes to fluke or accident.[15] But some intuitions of moral luck are due to undeserved credit or discredit that don't really have anything to do with flukes. That's because they involve things over which, let's say, the agent has no control and yet things that still obtain even in the relevant set of nearby possible worlds. This may be particularly true, for example, in cases of constitutive luck, or luck in character. A person's character may at least in part be due to his parents, and yet there is no possible world in which he has different parents. On this view of luck, then, much of what people term moral luck is actually moral *fortune*.[16] Consider another example, of someone who is an evil klutz, and who tries to harm people but instead ends up helping them. Let's assume this is part of his make-up, and in a wide class of the nearest possible worlds he is still an evil klutz, intending to harm but helping instead. Although the bad intentions, systematically across agents, produce bad outcomes, in his particular case they regularly do not. This evil klutz is not blamed to the extent that the competent evil person is. This is moral good fortune for the evil klutz. He has a bad character, of the sort that systematically produces bad outcomes in this world. He is deserving of blame for this, but not deserving of the same blame as the competent evil person who is actually harming others, and thus actually doing something wrong. One way in which we "lack control" is through accident or fluke. Another way is through simple lack of choice. These may or may not coincide.

So, we blame the morally unlucky because there is more to blame them *for*. This marks one difference between the two cases that can affect our intuitions about them. The truck driver who is reckless and the attempted murder both exemplify states of mind that typically do result in worse outcomes than good states of mind. In the case of the attempted murder, the murderer was morally lucky because in a wide class of the nearest possible worlds he succeeds, and he is a murderer rather than merely an attempted murderer. We can understand this also in terms of regularities for this world. In this world, when people intend to do bad things they are more likely to occur than otherwise. Bad intentions tend to generate bad outcomes. Blame is appropriate, then, for these states of mind even in the absence of bad outcomes in particular cases.

5 CONCLUSION

My claim is that outcomes matter to our evaluation of someone's action as right or wrong, although actions are only one of the things we blame and praise people for. A person's states of mind will matter when it comes to apportioning praise and blame to the person herself. But, again, they matter in a derivative way. They matter because they cause good outcomes. They cause such outcomes not infallibly, of course, but systematically or nonaccidentally.

In this paper my aim has been to show how such an account can accommodate luck and fortune in moral evaluation. Although the evaluational internalist has tried to avoid the problem by denying the relevance of effects to moral evaluation, this strategy cannot accommodate the deeply held intuition, voiced by Nagel, that the agent's actual impact on the world just does seem to matter to evaluation, that "what has been done, and what is morally judged, is partly determined by external factors" (Nagel 1979, 25). What this account of luck has tried to show is that within the objective consequentialist framework we can account for our ambivalence on this topic by embracing the nuanced forms of evaluation advocated by objective consequentialism.[17]

NOTES

1. Philosophers are also interested in luck in distribution of resources and burdens. This has nothing to do with moral evaluation. Someone could believe in the problem of moral luck, and yet also believe that luck in distribution of resources poses no problem. That is, it does not seem to pose the same sort of conceptual problem that moral luck poses. This is because we are used to thinking of the world as such that goods are not naturally distributed fairly. Luck in distribution of resources has to do with the fact that some people have more goods, or suffer more hardships, through good or bad luck in terms of their placement in society, or the sorts of opportunities that just happened to come their way in life. Some people are born poor, for example, and this is surely not their fault. They did nothing that would warrant poverty. Others are born wealthy, and for the same reason this seems undeserved—they did nothing to warrant their wealth. They were simply born into it. Even the exceptionally intelligent person who does work hard to get rich is also the recipient of luck, because his intelligence was something he was lucky enough to be born with. Of course people can work with and improve their natural talents and abilities, and improve the resources they began life with—but there is no denying that their starting point was the result of luck.
2. See, for example, Frances Howard-Snyder (1997, 241–248).
3. I discuss the theoretical problems for the mixed view in Driver (2001).
4. Michael Slote, for example, adopts this strategy for dealing with the solipsism problem for agent-based ethics in his (1997):

   > One doesn't count as genuinely benevolent if one isn't practically concerned to find out relevant facts about (certain) people's needs or desires and about what things are can or are likely to make them happy or unhappy. . . . One's inward gaze effectively "doubles back" on the world. (229)

5. Jennifer Lackey has recently attacked the "control" model of luck in her (2008). Pritchard (2005) also discusses problems with the "control" model. On the view I argue for here, lack of control is neither necessary nor sufficient for luck, but lack of control may be one of the set of pragmatic factors that goes into attributions of luck.

6. Some authors will view this as the same as contextualism, and that's fine with me. I don't want to get into a debate over the relative merits of contextualism and contrastivism and whether or not they are different or really the same thing.

7. See Jonathan Schaffer (2005).

8. See Walter Sinnott-Armstrong (2008).

9. See, for example, Pierre Laplace (1814).

10. See Rescher (1995, 35) for discussion.

11. Andrew Latus, mentioned in Pritchard (2005, 127).

12. Pritchard (2005, 126).

13. Ibid.

14. Pritchard (2005, 144n15) discusses such a possibility when examining Rescher's view.

15. Offhand it seems that flukes and accidents are not always the same. When a bird flies in front of the attempted murderer's bullet, that's a fluke and an accident, because unplanned. When a person wins a lottery, that's a fluke, but not an accident, because planned. When, in the attempted murder, the murderer trips and misses, that may not be a fluke, but it may be an accident, because unplanned.

16. I believe this gives us a way to solve at least one problem recently raised for the modal account. Jennifer Lackey presents the following counterexample to the simple modal account:

> BURIED TREASURE: Sophie, knowing that she had very little time left to live, wanted to bury a chest . . . on the island she inhabited. . . . Her central criteria were, first, that a suitable location must be on the northwest corner of the island—where she had spent many of her fondest moments in life—and, second, that it had to be a spot where rose bushes could flourish—since these were her favorite flowers. As it happens, there was only one particular patch of land on the northwest corner of the island where the soil was rich enough for roses to thrive. Sophie, being excellent at detecting such soil, immediately located this patch of land and buried her treasure, along with seeds for future roses to bloom, in the one and only spot that fulfilled her two criteria.
>
> One month later, Vincent, a distant neighbor of Sophie's, was driving in the northwest corner of the island—which was also his most beloved place to visit—and was looking for a place to plant a rose bush in memory of his mother who had died ten years earlier—since these were her favorite flowers. Being excellent at detecting the proper soil for rose bushes to thrive, he immediately located the same patch of land that Sophie had found one month earlier. As he began digging a hole for the bush, he was astonished to discover a buried treasure in the ground. (Lackey 2008, 261)

In this case, we have the intuition that Vincent is lucky, although it is also true that in nearby possible worlds he finds the treasure. What seems to be doing some of the work here is that Vincent could not reasonably foresee that he would find treasure in that spot, so he is quite surprised; the discovery is unplanned and utterly unexpected. Further, Sophie did not plan to leave it there for him to find, "fortuitously." There really was no plan at all for Vincent to find the treasure. It just worked out that way, but it did so in a way

reflected in nearby possible worlds. It was fortuitous but not flukish. This
should be treated as another good fortune case.

17. Work on this paper has been partly supported by a Fellowship from the
National Endowment for the Humanities. I would like to thank Walter Sinnott-Armstrong and Roy Sorensen for helpful conversations and comments
on this topic. An earlier version of this paper was presented at the first Online
Philosophy Conference, April 7, 2006. (experimentalphilosophy.typepad.
com/online_philosophy_conference).

I would like to thank Hans Maes for his extremely helpful comments
on the conference paper. The paper was also presented at North Carolina
State University, Raleigh, the University of Southern California, the University of Massachusetts at Amherst, the University of Miami, a workshop on
contrastivism at Dartmouth College, and at the Australian Association of
Philosophy Conference held at the University of New England, Armidale,
New South Wales, in July, 2007. I thank the members of those audiences
for their very helpful comments. I would particularly like to thank Richard Arneson, John Carroll, and Jonathan Schaffer for their detailed and
thoughtful comments.

REFERENCES

Driver, J. 2001. *Uneasy Virtue*. New York: Cambridge University Press.

Howard-Snyder, F. 1997. The Rejection of Objective Consequentialism. *Utilitas* 9:
241–248.

Lackey, J. 2008. What Luck is Not. *Australasian Journal of Philosophy* 86:
255–267.

LaPlace, Pierre. 1814. *Essai Philosophique sur les Probabilités*. Paris: Mme. Ve.
Courcier.

Nagel, T. 1979. *Mortal Questions*. New York: Cambridge University Press.

Pritchard, D. 2005. *Epistemic Luck*. Oxford University Press.

Rescher, N. 1990. Luck. Presidential Address before the Eighty-Fifth Annual East-
ern Division Meeting of the American Philosophical Association, 1989, Atlanta,
GA. In the *Proceedings and Addresses of the American Philosophical Association* (1990) 64: 5–19.

Rescher, N. 1995. *Luck: the Brilliant Randomness of Everyday Life*. New York:
Farrar, Strauss, & Giroux.

Schaffer, J. 2005. Contrastive Knowledge. In *Oxford Studies in Epistemology 1*,
ed. T. Gendler and J. Hawthorne. Oxford: Oxford University Press, 235–271.

Sinnott-Armstrong, W. 2008. A Contrastivist Manifesto. *Social Epistemology*
22(3): 257–270.

Slote, M. 1997. Virtue Ethics. In *Three Methods of Ethics*, ed. M. Baron, P. Pettit,
and M. Slote. Malden, MA: Blackwell Publishers, 175–238.

Williams, B. 1981. *Moral Luck*. Cambridge: Cambridge University Press.

Contributors

Martijn Blaauw is Assistant Director of the 3TU.Centre for Ethics and Technology, and a senior research fellow at Delft University of Technology. He has published widely in epistemology, including recent publications in such journals as *Analysis* and *Australasian Journal of Philosophy*. He is currently trying to defend a particular account of the notion of 'privacy' in terms of knowledge.

Julia Driver is Professor of Philosophy at Washington University in St. Louis. Her research interests lie primarily in normative ethical theory and moral psychology. She is the author of articles appearing in journals such as *Ethics, Nous, Hypatia, Philosophy, Philosophy & Phenomenological Research, Philosophical Studies,* and the *Journal of Philosophy*. She is the author of *Uneasy Virtue* (Cambridge University Press, 2001), *Ethics: the Fundamentals* (Blackwell, 2007), and *Consequentialism* (Routledge, 2012).

Branden Fitelson is currently Associate Professor of Philosophy at Rutgers University and visiting Professor of Philosophy at the Munich Center for Mathematical Philosophy at the Ludwig-Maximilians-University (LMU) in Munich. Before teaching at Rutgers and LMU, Branden held teaching positions at UC-Berkeley, San José State, and Stanford. Branden got is MA and PhD in philosophy from the University of Wisconsisn-Madison. Before getting into philosophy, Branden studied math and physics at UW-Madison, and he worked as a research scientist at NASA.

Christopher Hitchcock is Professor of Philosophy at the California Institute of Technology. He is the author of numerous articles on causation, explanation, and other topics in the philosophy of science, which have appeared in most of the leading journals in philosophy and philosophy of science, as well as in a number of distinguished edited collections. Three of his papers have been selected for inclusion in the Philosopher's Annual. He has served on the governing board of the Philosophy of Science Association, and as president of the Society for Exact Philosophy.

Adam Morton is at the University of British Columbia. He has two forthcoming books, *Bounded Thinking* (Oxford), and *Emotion and Imagination* (Polity).

Jonathan Schaffer is a professor of philosophy at Rutgers University. He has previously held positions at the Australian National University, University of Massachusetts-Amherst, and University of Houston. His work has received several awards, including a selection in Philosopher's Annual (2009), the American Philosophical Association's Article Prize (2008), and the Young Epistemologist Prize (2002).

Walter Sinnott-Armstrong is Chauncey Stillman Professor of Practical Ethics in the Philosophy Department and the Kenan Institute for Ethics at Duke University. He received his BA from Amherst College in 1977 and his Ph.D. from Yale University in 1982 then taught at Dartmouth College until 2009. He has served as vice-chair of the Board of Officers of the American Philosophical Association and co-director of the MacArthur Project on Law and Neuroscience. He publishes widely in normative moral theory, meta-ethics, applied ethics, moral psychology and neuroscience, philosophy of law, epistemology, informal logic, and philosophy of religion. He is a committed contrastivist.

Justin Snedegar is a PhD candidate at the University of Southern California who works in ethical theory, especially metaethics and practical reasoning. His paper 'One Ought too Many', co-authored with Stephen Finlay, is forthcoming in *Philosophy and Phenomenological Research*.

Index

Note: Page numbers followed by "ff" indicate the page and subsequent pages that follow.